SMALL LUXURIES

FILM CAMERAS

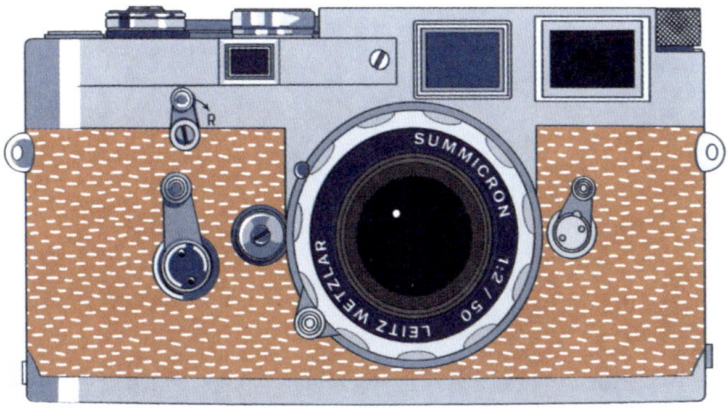

SMALL LUXURIES

FILM CAMERAS

Aaron Lurth

motorbooks

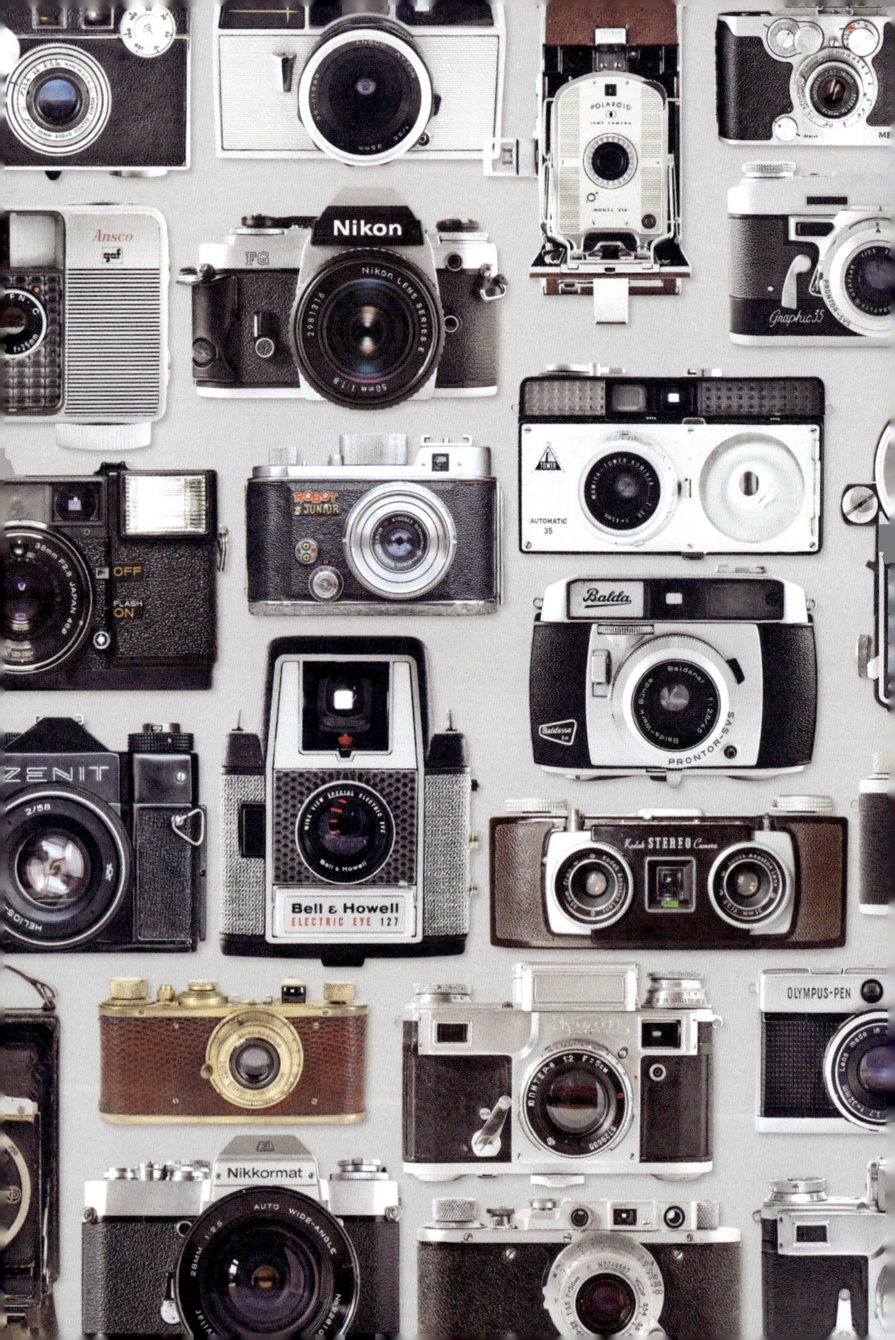
Nikon
FG
Nikon LENS SERIES E
Ansco
gaf
POLAROID
Graphic35
ROBOT
JUNIOR
AUTOMATIC
35
OFF
FLASH
ON
Balda
PRONTOR-SVS
ZENIT
Bell & Howell
ELECTRIC EYE 127
Kodak STEREO Camera
OLYMPUS-PEN
Nikkormat
AUTO WIDE-ANGLE

CONTENTS

CHAPTER ONE

The Alchemy of Light and Silver

A Brief History of Photography

IN THE TIME IT TAKES YOU to finish reading this sentence, almost half a million new photographs will be taken worldwide. Every day, five billion new images are created, a pace that will produce a little over one trillion images this year alone. These are staggeringly large numbers, but volume alone cannot define the impact photography has had on the world.

Photographs have introduced us to events, places, and people that most of us could never experience on our own. Images have recorded massive celestial objects in deep space light-years away, and they've revealed objects smaller than the tip of a needle. They have recorded realities we didn't know existed and immortalized fleeting moments of our everyday lives.

Photography has woven itself into the fabric of our lives so thoroughly that it's hard to imagine a time when it didn't exist. In spite of its ubiquity in modern society, though, photography is a relatively young medium, one with a fascinating past.

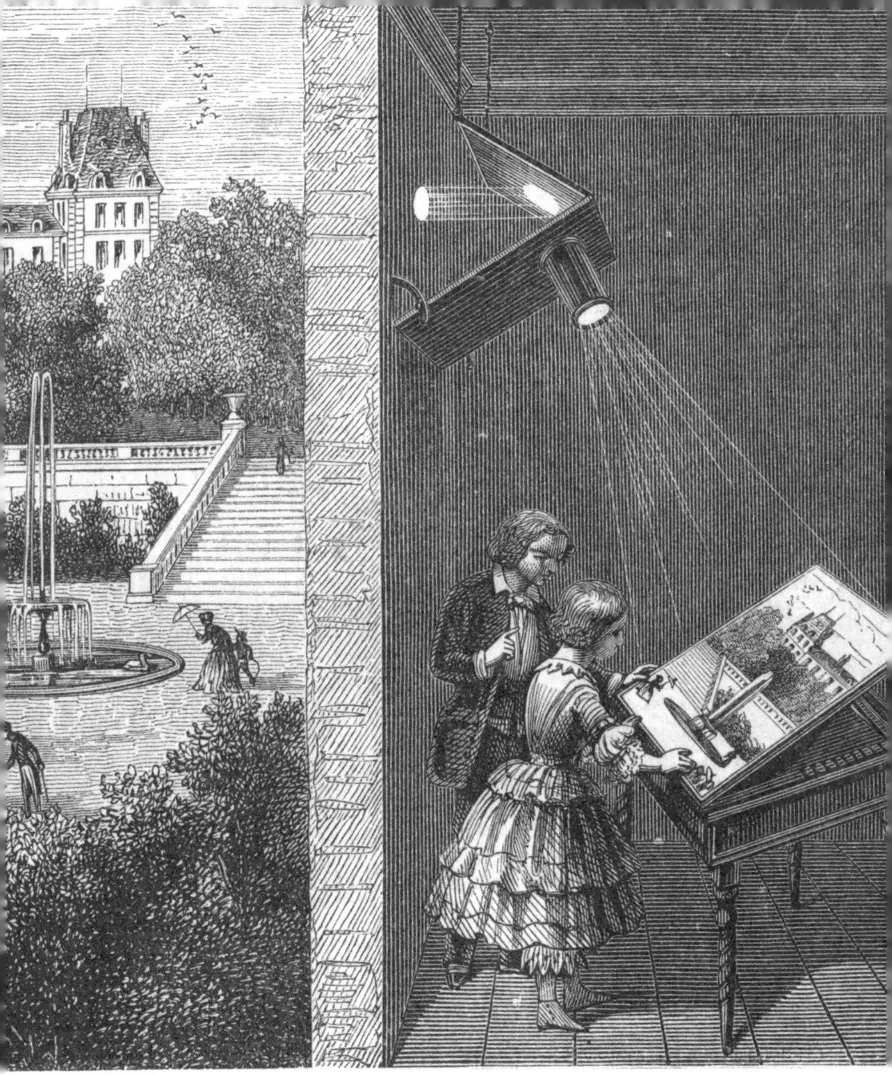

Using the same principles of physics that make creating permanent images with cameras possible, camera obscuras allowed viewers to experience projections of images into rooms hundreds of years before the handheld camera was invented.

The Fleeting Images of the Camera Obscura

Long before chemists discovered how to permanently record images on pieces of paper, early observers of optics noted that they could project inverted images of scenes onto one wall of a darkened box when they focused light through a small hole on the opposite side of the box. The phenomenon was first documented in 400 BCE by the Chinese philosopher Mo-Tzu. While the concept was widely known for more than a millennium, the camera obscura—a functional box for projecting images—wasn't invented until the eleventh century.

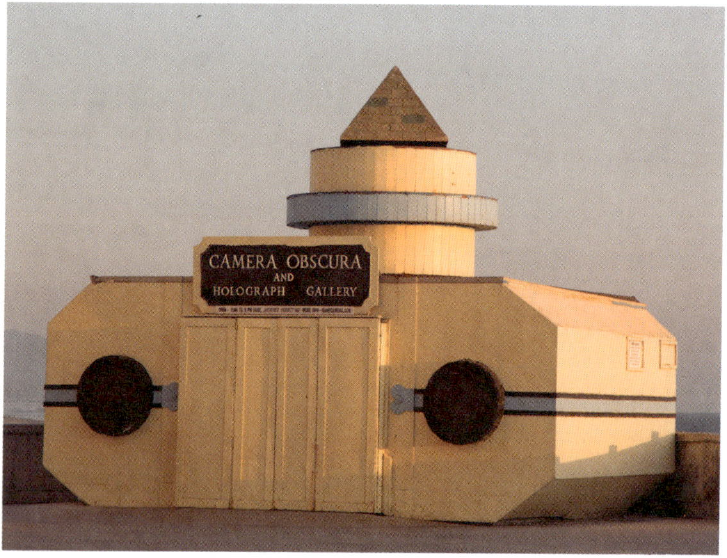

Built in 1946, this camera obscura, located on the grounds of San Francisco's Cliff House in the Golden Gate National Recreation Area, is an example of historic camera obscuras built as seaside attractions.

Working in Cairo in the first half of that century, the Arab scientist, astronomer, and mathematician Hasan Ibn al-Haytham (known in the West as Alhazen) performed his experiments in a darkened room with a small hole that let in light. Alhazen set up a projection screen inside the room on which he projected images from outside. Through several experiments involving a series of lamps set outside his early version of a camera obscura, Alhazen observed that light traveled in a straight line. He used this principle to explain why the images were always projected upside-down and backward.

The camera obscura evolved over the next seven hundred years, with additions such as lenses to focus and sharpen the projected image and mirrors to flip the images right-side up. The camera obscura became smaller, lighter, and more mobile—eventually constructed as a small tent that could be set up anywhere. Strong evidence suggests that famous artists such as Rembrandt, Canaletto, and even Vermeer used camera obscuras as drawing and painting aids, allowing them to trace projected images on canvas.

The Chemistry of Freezing Time

While new discoveries in physics, optics, and engineering helped advance the quality of the camera obscura, chemistry offered the missing piece that finally made photography possible. The term *camera obscura* first appeared in a 1604 work of the astronomer Johannes Kepler, where he described the "darkened chamber" he used to view sunspots. The device was more often featured as a novelty whose magic delighted viewers—mostly in seaside attractions or traveling shows—but its transient images left some wanting more. Scientists, inventors, and artists sought ways to permanently fix those projected images in place. The breakthrough came in the first half of the 1800s, when a number of scientists and artists discovered solutions to the problem at virtually the same time. With these early attempts to capture images, photography as we know it was born.

The answer lay in chemistry. Early scientists had noted that silver salt solutions darkened when exposed to light. Over one hundred years before the first actual, permanent photo was taken, the German doctor Johann Heinrich Schulze discovered that a solution he had left in a window had turned deep purple. That solution—a mixture of chalk, silver, and nitric acid—was sensitive to light. Starting in 1717, Schulze performed a number of primitive experiments, covering various parts of a glass bottle with letter stencils to produce photographic messages for his friends. These eventually faded, though, as the solution continued to react to light and change color. The essential ingredient for making a light-sensitive material was silver, which would be the key to creating most of the analog photographic processes we know today.

More than a century later, in 1827, Joseph Nicéphore Niépce coated a highly polished pewter plate with oil of lavender infused with bitumen of

View from the Window at Le Gras, *taken by Nicéphore Niépce in 1827, is the oldest surviving photograph. The image was captured using Niepce's heliograph process and a camera obscura for the camera.*

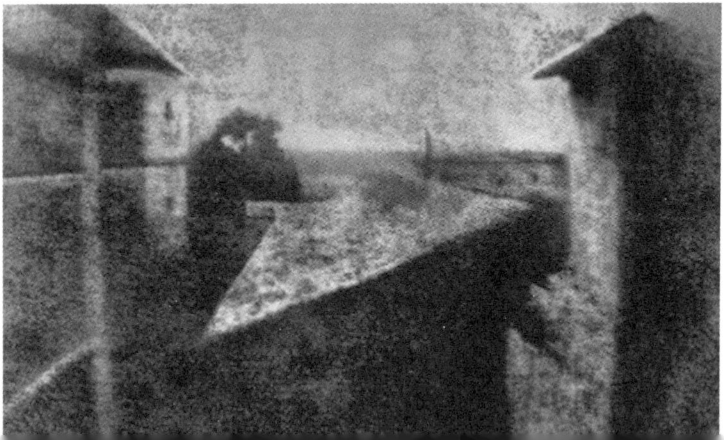

Judea and exposed it to light coming through his attic window. A naturally occurring asphalt, bitumen of Judea hardens when exposed to light. After developing the plate by washing away the underexposed portions with turpentine, Niépce was left with the first permanent photographic image—and the oldest surviving photograph on record. Dubbed "heliography," this technique had limited uses for photographic reproduction, as it required an extraordinarily long exposure time (eight hours, in this case). The invention of a commercially viable photographic process lay elsewhere, though Niépce contributed to its discovery.

While Niépce busily experimented with his heliograph process, some 220 miles (350 kilometers) away, Louis Daguerre was running a diorama theater in Paris. An artist and showman, Daguerre featured a 45×69-foot (14×21-meter) painting illuminated by various light sources, affording spectators a lifelike viewing experience. Daguerre often used a camera obscura to help create his large dioramas, but he wanted a way to permanently record his projections. Eventually Daguerre and Niépce crossed paths, collaborating on experiments with fixing images in place.

Another development occurred in England, where inventor William Henry Fox Talbot was working on his own process. By brushing writing paper coated in salt with silver nitrate, Talbot found he could make permanent images on paper. He made precise reproductions of botanicals—which he called "drawings"—by pressing leaves and plants between his coated paper and a sheet of glass. He then laid the stack of paper, glass, and botanicals in the sun. As the light hit the paper, the fully exposed areas became dark, effectively making a negative image of the botanicals. It was a highly effective and relatively easy process, but one Talbot kept to himself for years.

After Niépce's death in 1833, Daguerre continued his experiments alone and eventually stumbled upon his photographic process through a happy accident. Having experimented with highly polished, silver-coated copper plates, Daguerre and Niépce were at a loss to make a latent image appear on the plate's surface after exposing it to light.

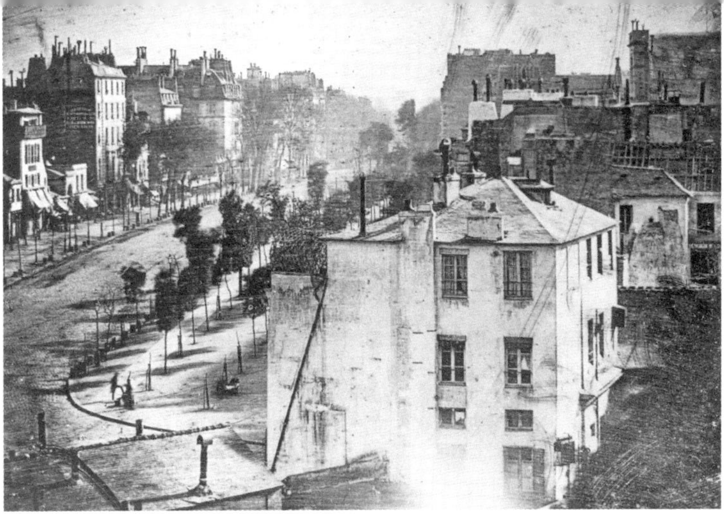

Boulevard du Temple, *taken by Louis Daguerre in 1838 in Paris, is an example of an early daguerreotype. This image is considered one of the first images of a candid photograph of a person.*

One morning, Daguerre returned to a failed plate he'd set in a chemical cabinet the night before and discovered a preserved image on the plate. Through trial and error, he determined that the helpful culprit was an open bottle of mercury: mercury vapor was the key. The results were hauntingly beautiful, metallic images that had an almost three-dimensional quality—a mirror with a memory. He named the process that created these early photographs after himself: daguerreotype.

Daguerre and Talbot could be considered the godfathers of photography, but Daguerre's ruthless self-promotion served to eclipse Talbot's efforts. By 1841, when Talbot perfected his paper negative process and introduced "calotypes" to the world, the daguerreotype process was already in wide use and would eventually win out as the standard photographic process for a time.

Photography was still evolving, however, and the daguerreotype couldn't hold its place as new, more efficient methods appeared. The daguerreotype process had serious drawbacks—it relied on highly toxic substances, and the one-off images it created could not be reproduced—and it was eventually replaced by the wet-plate collodion and dry-plate silver gelatin processes.

The wet-plate collodion process had a few distinct advantages over daguerreotypes. It didn't require mercury vapor. Exposure times were mere seconds in bright sunlight, or a couple of minutes indoors, making this technique much faster than producing daguerreotypes. And the negatives produced could be used to make as many positive copies as desired, allowing for mass-produced images.

This process came to dominate the photographic landscape in the late 1800s. Photographers such as Mathew Brady and Alexander Gardner used the method to document the American Civil War. Images from the Crimean War taken by Roger Fenton were also created with collodion plates. And the images taken by William Henry Jackson for Ferdinand

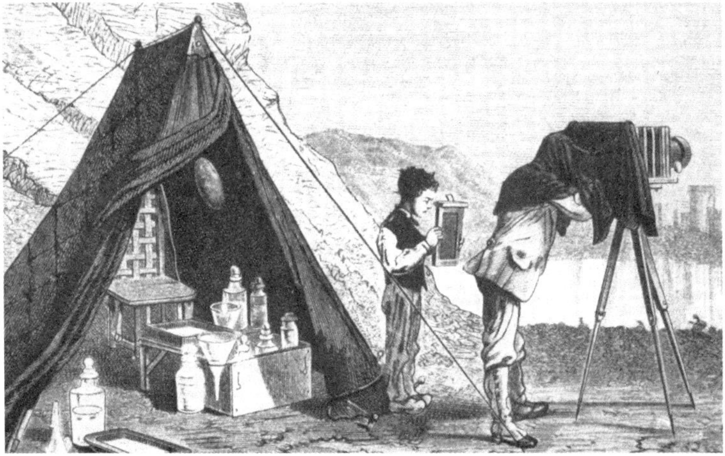

The wet-plate collodion process, popular in the late 1800s, required the photographer to travel with a portable darkroom to prepare the photographic plates on the spot.

Hayden's surveys of the American West produced such jaw-dropping images of Montana that they moved the United States Congress to create Yellowstone National Park, the first national park in the country.

The wet-plate collodion process wasn't perfect, though, and certainly not convenient. It required the photographer to carry a portable darkroom, essentially an entire field laboratory, and the plates had to be prepared on the spot using different liquid chemicals. When the subject was ready to be photographed, the plate needed to be exposed while wet and developed immediately. Between prepping the plate, taking the photo, and developing the image, the photographer had to work fast to avoid spoiling his work.

The entire process saw drastic improvement with a new discovery: By suspending photosensitive silver nitrate in a gelatin solution and allowing it to dry on a substrate, plate exposure times were significantly reduced. The negatives produced in this way could be stored and used later. The result was a shelf-stable emulsion that could be applied to glass as well as many other surfaces.

The Democratization of Photography

The next evolutionary step took photography out of the hands of experts and brought it to the masses. Eastman Kodak Company's Kodak No. 1 in 1888, followed two years later by the Kodak Brownie camera, made it possible for anyone to capture and preserve images. Kodak sought to make image taking as easy as possible for the consumer, producing small, portable cameras that came preloaded with film. When you shot through the entire roll, you simply sent the camera back to Kodak, who developed and printed the photos, loaded a new roll of film, and sent the camera and printed photographs back to you. The company slogan, "You press the button, we do the rest," perfectly defined their business model and helped break down the barriers of entry into the world of photography. The cameras were a smash hit: In the first year alone, Kodak's one-dollar Brownie camera sold over 150,000 units.

Led by Kodak, the invention of the film roll ushered in a new era of camera design. Cameras could now be compact and portable. The company created a numbering system for film sizes, the same system used today. Based on the order the film was created, not specifically on film sizes, the numbers allowed for simple identification of the film you were shooting and industry standardization of film size. While the list of sizes is large, the most commercially successful sizes included 120 and 220 (known as medium format), 135 (commonly referred to as 35mm), 127 (a square-format, Vest Pocket film, as Kodak would call it), and 110 (miniature cartridge film).

The German company Leica was the first to introduce a commercially successful 35mm film camera, the Leica 1, in 1925. Striking the perfect balance between compact size, image quality, and practicality, 35mm went on to dominate the film photography landscape. Most of the consumer cameras sold in that time were 35mm cameras, while larger formats were typically used in commercial and scientific settings or by serious hobbyists.

While 35mm remained the de facto standard for the general public, another film process enjoyed

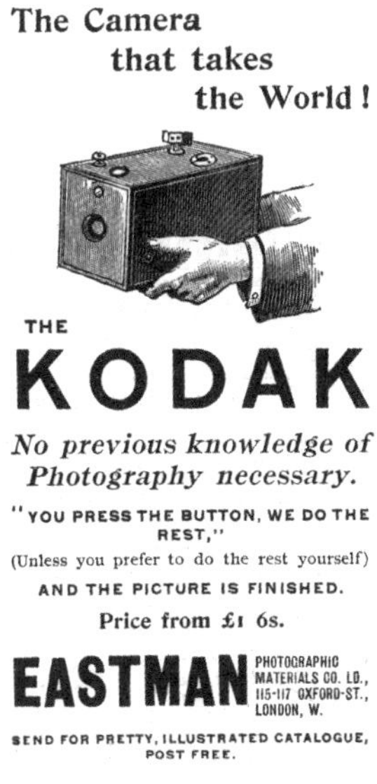

With the slogan "You press the button, we do the rest," Kodak removed the barrier of entry to photography, no longer requiring the knowledge of chemistry or camera operation.

popularity in the second half of the twentieth century. This was the Polaroid instant camera, debuted in 1948 by Edwin Land. Employing a novel formula, Polaroid film contained all the chemicals needed to develop a photograph on the spot. As the technology progressed, color photographs became available to an even greater population. Development time dropped to a minute, allowing for images to appear magically right before your eyes. At its peak in the late 1970s, Polaroid controlled two-thirds of the instant photo market.

Where there were profits to be made, so followed rapid innovation and development. Cameras of all shapes and sizes emerged from the 1900s through the early 2000s, from rangefinder cameras (like many of the classic Leicas) to single-lens reflex cameras (SLRs) and compact point-and-shoots. Eventually the addition of electronic circuit boards and motors gave way to a whole new generation of "cameras with a brain." What started as a way to assist photographers with electronic light meters and advancing and winding of film led to automatic exposure modes, autofocus (AF) cameras with intelligent subject tracking, and even novel functions like Canon's eye-controlled autofocus system. Suddenly, cameras were smarter and easier to operate than ever before.

Originally debuted in the EOS 5, Canon's novel eye-controlled autofocus system could track the photographer's eye and focus wherever they were looking in the viewfinder.

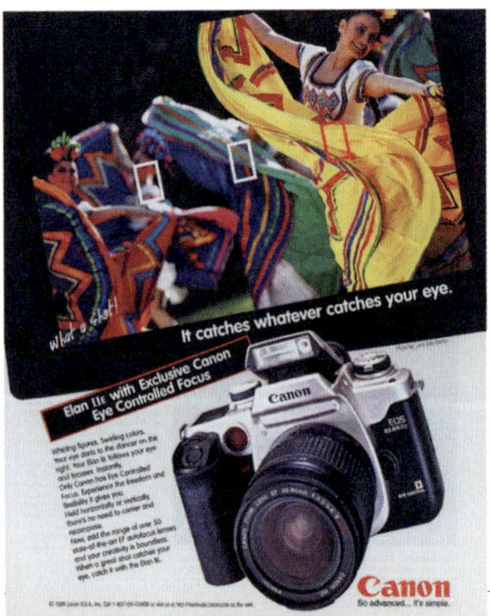

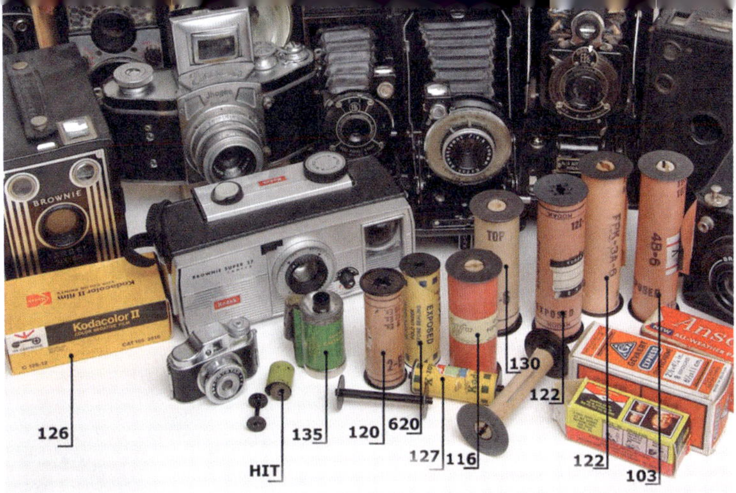

While 35mm and 120mm would be the most commercially successful film sizes, a number of different roll sizes were created and explored throughout the early years of photography.

The Fall from the Mountaintop

Film cameras increased in popularity, and sales rose steadily, from the 1950s until the 1980s. That decade, sales saw a dramatic increase, with the peak in cameras sales at about 37 million units in 1998. During the 1990s, an estimated 90 percent of all American households owned a camera. Entire economies popped up around film camera usage, from local camera clubs, camera stores, and repair shops to one-hour photo-developing centers. Life was great in the film camera world. Until it wasn't.

Cameras had evolved quickly since the early days of photography, and they had exploded in popularity by the end of the twentieth century. That ended in the early 2000s, as film cameras quickly fell out of favor among the general public and were replaced by digital cameras. Worldwide demand for photographic film by 2010 was down to just 10 percent of its sales peak only twelve years earlier, and a few years later nearly all major camera manufacturers halted production of new film cameras entirely.

While sales in the early 2000s looked grim, film photography wasn't dead. The 2020s have brought a resurgence of interest in the medium. Driven in part by younger generations who didn't grow up using film cameras—along with trends set by celebrities and social media influencers—film and film camera sales began to rise again. One illustration of this is the release in 2023 of a brand-new film Harman Phoenix by Harman Technologies (Ilford's parent company), with a second film camera appearing in 2025, the Red 125. In 2024, a year after the Phoenix appeared, Pentax surprised the world with a new half-frame point-and-shoot camera (two pictures on one 35mm frame) called the Pentax 17, the first new production film camera from a major camera brand in twenty years.

If the film revival has shown us anything, it's that, while industries and technologies change, the same magic of film cameras that captivated people one hundred years ago persists today—and will continue into the future.

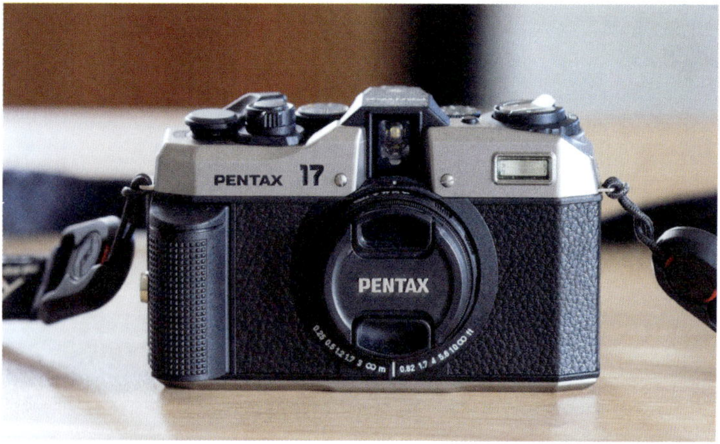

Released in 2024, the Pentax 17 is a half-frame camera, able to take seventy-two images on a thirty-six-exposure roll of film.

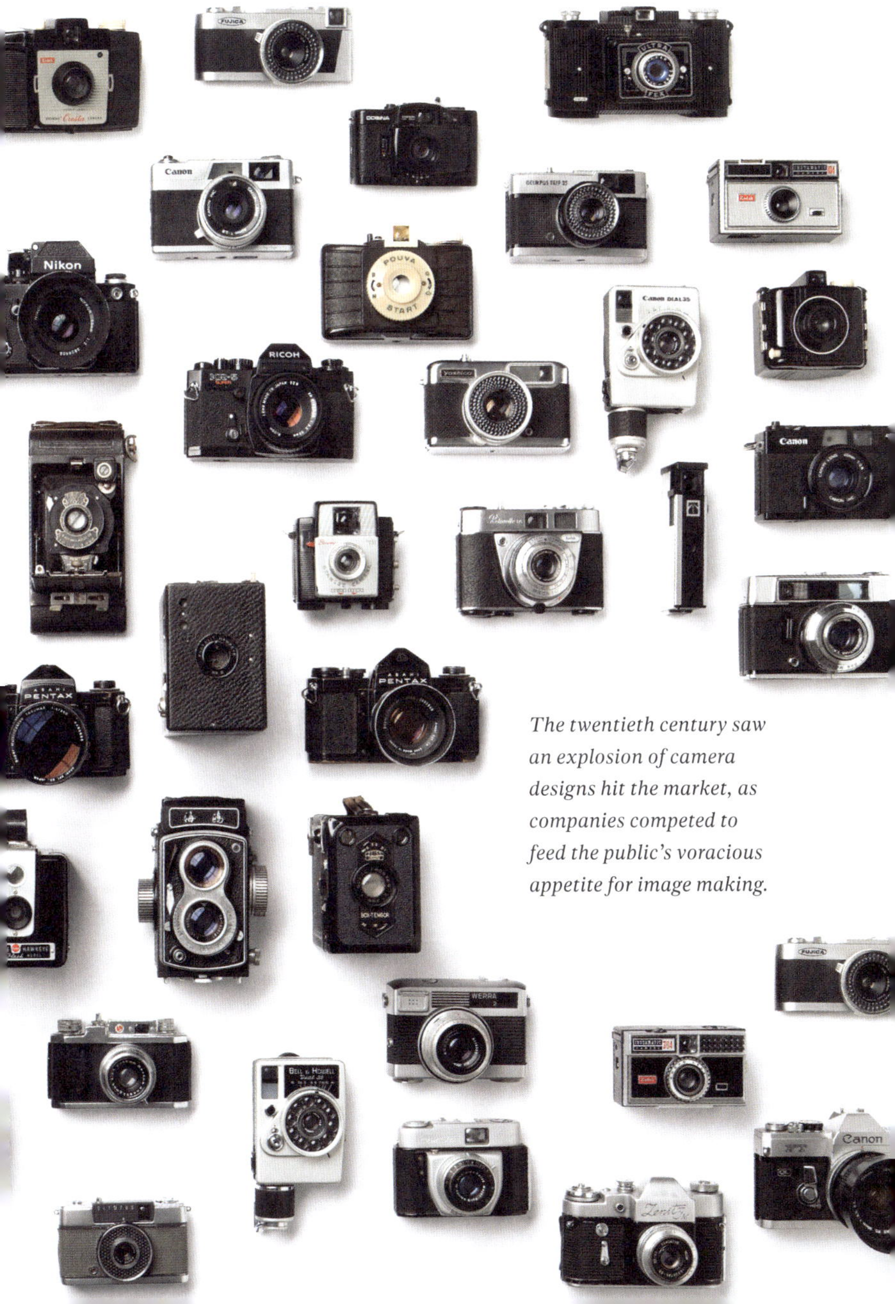

The twentieth century saw an explosion of camera designs hit the market, as companies competed to feed the public's voracious appetite for image making.

CHAPTER TWO

Inside the Black Box

How a Camera Works

FOR OVER A CENTURY, film cameras have represented the pinnacle of precision manufacturing, advanced optical and mechanical engineering, and creative industrial design. For anyone who's taken a moment to look through the many examples of old film cameras, the many creative ways these devices were designed becomes instantly apparent. When stripped of all the unique construction, creative designs, and complex mechanical mechanisms, though, all cameras are essentially the same: light-tight devices with a small opening on one side and, typically (though not always), a photosensitive material on the opposite side.

The classic Rolleiflex is a prime example of the innovative mechanical engineering and inspired design that have been hallmarks of the most popular film cameras over the decades.

The camera's functionality is entirely based on the physical properties of light. When light rays from various external points pass through a small opening in the darkened space, the light rays become focused enough to project an image of the external

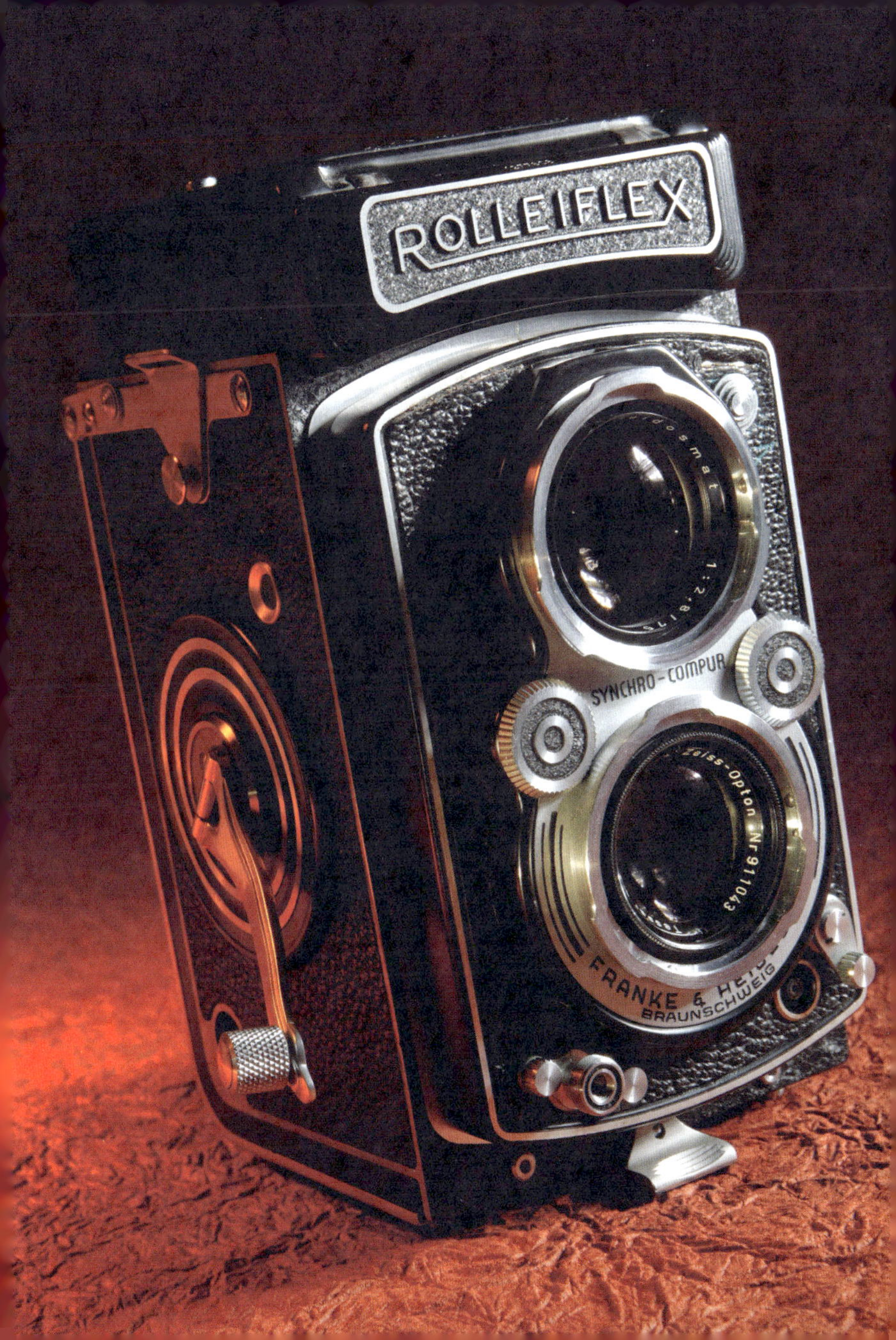
ROLLEIFLEX
SYNCHRO-COMPUR
Zeiss-Opton Nr 911043
BRAUNSCHWEIG

scene onto whatever surface they strike in that space. Since light travels in a straight line, focused light rays pass through the small opening and continue to travel straight from their origin. The resulting projected image is both upside-down and flipped horizontally. This means that if you stand on a hill and set up a simple camera obscura to view a busy street scene below, the cars will be upside-down at the top of your projected image and the sky will be at the bottom.

Our own eyes function in much the same way: The fluid-filled chambers in our eyeballs act as the space into which light passes, first through the cornea, then through a small opening (the iris) into that darkened space. The light is then focused through a lens and an image projected onto photosensitive material on the opposite side of the opening (our rods and cones). The image is converted into electrical signals sent to the brain, which interprets what has been seen. The image projected into our eyes is also upside-down, just as it is in a camera. Our brains flip this image right-side up to correct for this aberration.

What's cool about this phenomenon is that the darkened space needed to create a camera can be anything: an empty soda can, a cereal box, a trash can, an airplane hangar . . . all have been used as rudimentary pinhole cameras in the past.

Aperture

The small opening in the front of a camera is the aperture. While the aperture and the shutter are both responsible for controlling the amount of light coming into a camera, the aperture does so by varying the size of the opening into the camera. The aperture doesn't need to be complicated. In the case of a pinhole camera, for example, the aperture is simply a small hole on any light-tight object; the tiny pinhole

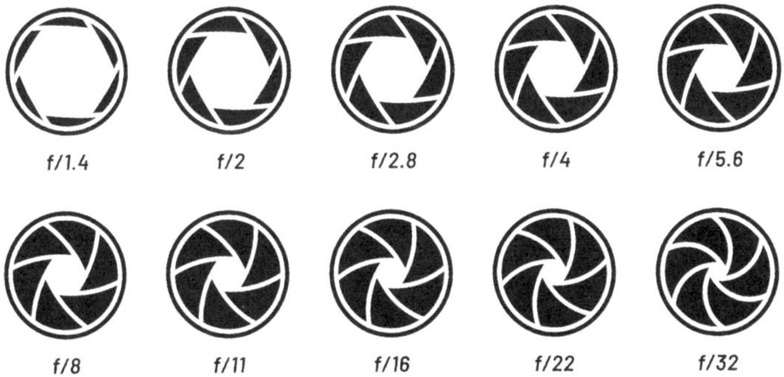

The aperture setting is indicated by a fraction that divides the lens focal length by the size of the aperture opening. Each time you move up or down one full stop, you either increase or decrease, respectively, the amount of light entering the camera by double or half.

is small enough to focus light through its opening and project an image onto the surface of a light-sensitive material. When we think of the opening used in actual film cameras, the aperture is a little more complicated.

A circular, mechanical aperture is what most of us think of when we refer to a camera's aperture. This is typically found inside of the lens and can reasonably be compared to the iris in the human eye. It consists of multiple, overlapping blades that form a circle. Just like our iris, the aperture can expand and contract, growing or shrinking the size of the lens opening—effectively letting in more or less light.

The size of the aperture is called an f-stop, with each f-stop assigned a number. Before the advent of mechanical, overlapping blades to adjust the opening inside the lens, metal plates with holes of varying sizes could be inserted into the lens to adjust the aperture. These plates, known as "Waterhouse stops," were inserted into lenses of specific focal lengths. Eventually the concept of the "stop" was combined with "focal" and abbreviated to the term we use today: f-stop.

The numbers used for aperture settings are fractions derived from taking the lens focal length and dividing it by the size of the aperture opening. Common full stops often appear as the following: f/1, f/1.4, f/2, f/2.8, f/4, f/5.6, f/8, f/11, f/16, f/22, and f/32. While it may not look like it, each time you move up or down one full stop, you either increase or decrease the amount of light coming into the camera by double or

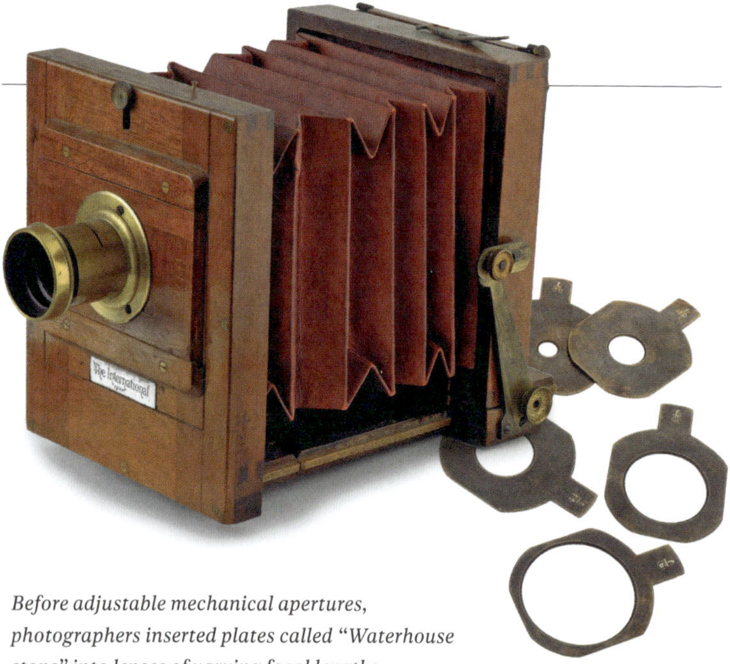

Before adjustable mechanical apertures, photographers inserted plates called "Waterhouse stops" into lenses of varying focal lengths.

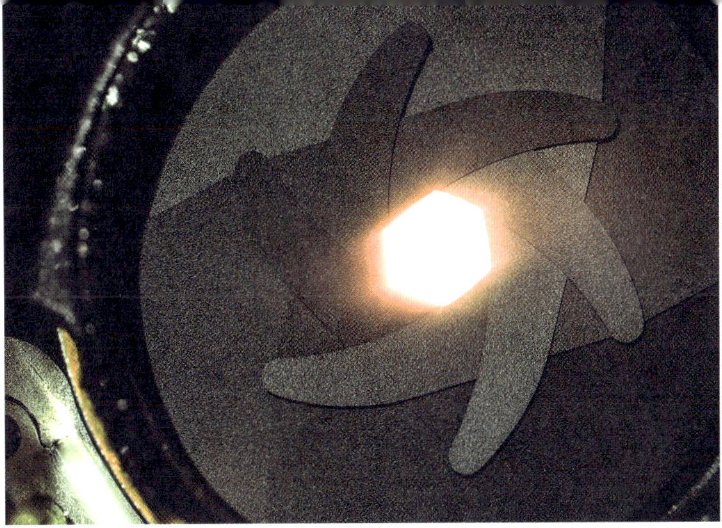

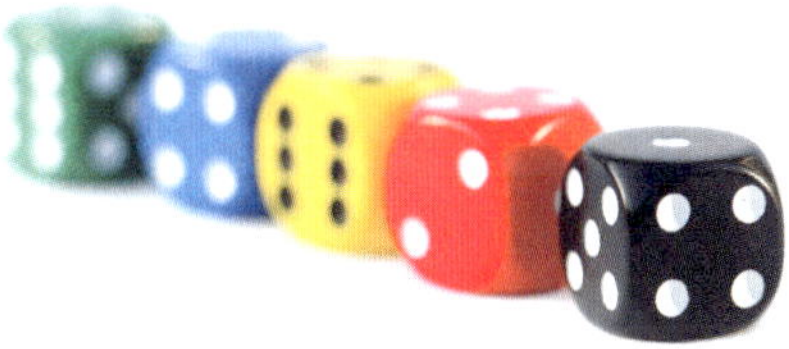

Above: Mechanical, overlapping blades allow the photographer to adjust how much light enters the camera through the aperture.
Left: Changing the f-stop affects what is in and out of focus in front of and behind the object you're focused on. This resulting sharpness is referred to as "depth of field."

half. For example, f/2.8 lets in twice as much light as f/4. Replacing the "f" with a number one helps make it clearer: 1/2 is indeed bigger than 1/4.

In addition to determining the amount of light entering the camera, the aperture has another effect on the image: Changing the f-stop will change the focus on objects in front of and behind what you're focusing on. This degree of sharpness is referred to as depth of field. As the aperture gets smaller—for example, moving from f/2.8 to f/11—the depth of field increases and more of the scene becomes sharp and into focus.

MAKE A CAMERA OBSCURA

Did you know you can turn virtually any room into a camera? A camera obscura (literally "dark chamber") is any device that projects an image through a small hole into a darkened space and onto the opposite side of that space. Pinhole cameras are a kind of camera obscura.

For this experiment you'll need a few basic items:

- Heavy-duty black trash bags (as many as necessary to cover your windows)
- Scissors
- Painter's or duct tape
- A small room with a window

To create your camera obscura, you need a bright, sunny day and a room that you can make totally dark. Take the trash bags and cut two of the sides to make each a single, flat sheet of plastic.

Cover your windows with the cut trash bags, taping them so that no light comes through. If you still see light coming through the bags, you may need to add a second layer.

By fully covering your windows, you've created one of the key elements of a camera: a darkened space.

For the second element, you need a small opening—that is, an aperture. Find a point somewhere in the middle of the covered windows and cut a hole in the trash bags. The size of your aperture will affect the brightness and sharpness of the image projected into the room. Experiment with the size of the opening, but start small (0.5–1 inch/1.25–2.5 centimeters in diameter) and slowly increase the size from there.

Once the aperture has been cut, you should see an image of the outside world projected into the room. You've just created a rudimentary camera!

A film camera does two things, which can be summed up quite simply: It controls the amount of light allowed into a light-tight device and how long that light is allowed to enter. These two functions are controlled by two different operations: the aperture and the shutter. Next we'll see how each works and how they affect the qualities of the final image when taking a photograph.

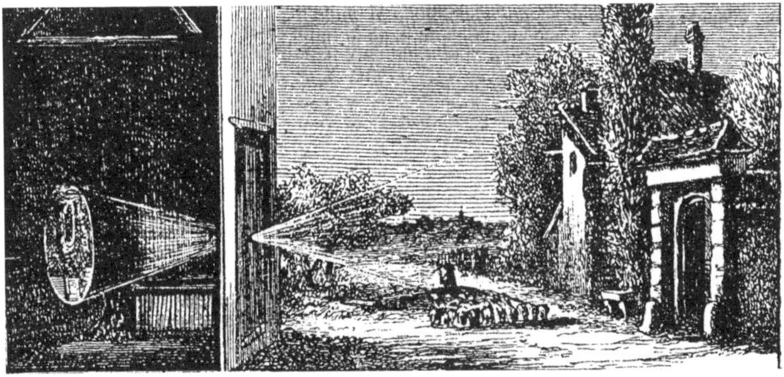

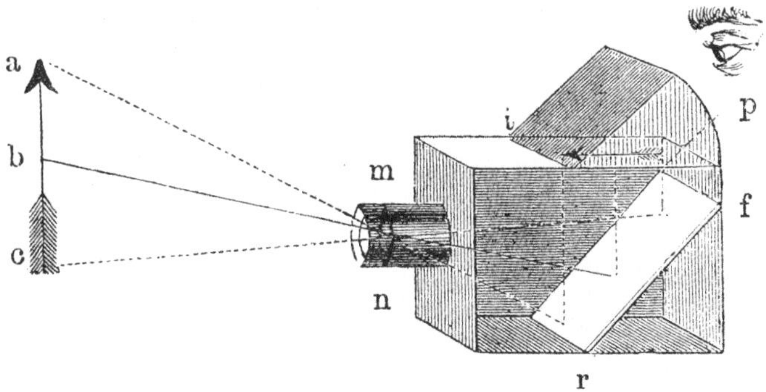

This illustration of a camera obscura illustrates how any darkened space—even a room—with a hole for light to enter can be turned into a camera.

Shutter

While aperture controls how much light can enter the camera by changing the size of its opening to the outside world, the shutter is responsible for determining the amount of *time* the light is allowed in. Film cameras can have either a focal-plane shutter or a leaf shutter.

Focal-plane shutters are the most common type of shutter found in single-lens reflex (SLR) cameras (see chapter 3). They can also be found in other types of cameras, including some large-format cameras like the

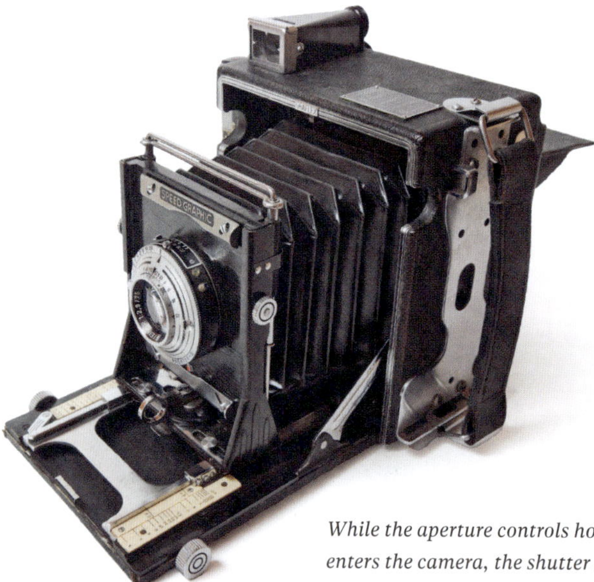

While the aperture controls how much *light enters the camera, the shutter determines* how long *the light is allowed into the camera. The Graflex Speed Graphic is a unique example of a camera that contains both a focal plane and leaf shutter.*

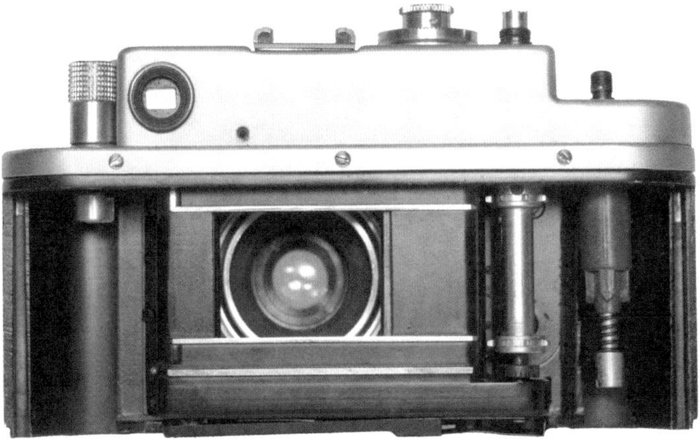

A focal-plane shutter comprises a pair of curtains that allow light to travel past and reach the film.

Graflex Speed Graphic. The focal-plane shutter consists of a pair of curtains located inside the camera, sitting in front of the film. During exposure, the curtains form a slit that travels across the film plane and exposes the film to light. The size of the slit can be adjusted, which in turn determines the amount of time (speed) the film is exposed to light.

A leaf shutter is usually built into the camera lens, rather than in the camera body itself. These shutters consist of leaves (or blades) that, in older cameras, act as both the aperture and the shutter. Instead of forming a slit that travels across the film plane, leaf shutters open from the center, expanding outward and stopping at the selected aperture size, then close back down. Leaf shutters are not as fast as focal-plane shutters, but they offer flash synchronization at every shutter speed and don't suffer from "rolling shutter" effects that can occur with focal-plane shutters.

Shutter speeds are typically listed as fractions. As with aperture stops, moving up or down in (full) shutter speed stops will double or halve the amount of light coming in. For example, going from a shutter speed of 1/30 (of a second) up to 1/60 means the shutter will be open for half the amount of time and half the amount of light will be let into the camera.

Because the shutter is responsible for the amount of time the light strikes the film, the shutter is also responsible for slowing or freezing motion in the final image. If an object in the image is moving fast enough to travel across the film while the shutter is open, that object will create motion blur in the final image. Conversely, if the shutter travels faster than an object is moving, that object will appear as frozen in place in the final image. Modern shutter speeds are sometimes thought of as taking a "scalpel to time," as many of the higher-end film cameras can shoot at shutter speeds as fast as 1/8000 of a second.

Putting It All Together

To capture a well-exposed image, the film you select needs a certain amount of light, known as exposure. Think of exposure as an equation: shutter speed + aperture = exposure. In most cases, to obtain a perfectly exposed image the equation should equal 100 percent. That ratio and the total amount of light required will be different for each picture you take, but most cameras have a built-in light meter to help show you how much light is needed to make a well-exposed image.

A photographer needs to decide on a number of factors when taking a photograph. Two key questions are:

- Do you need to freeze motion or show motion?
  (The answer will determine the required shutter speed range.)
- How much of your photo do you want to be in focus?
  (This will determine the required aperture range.)

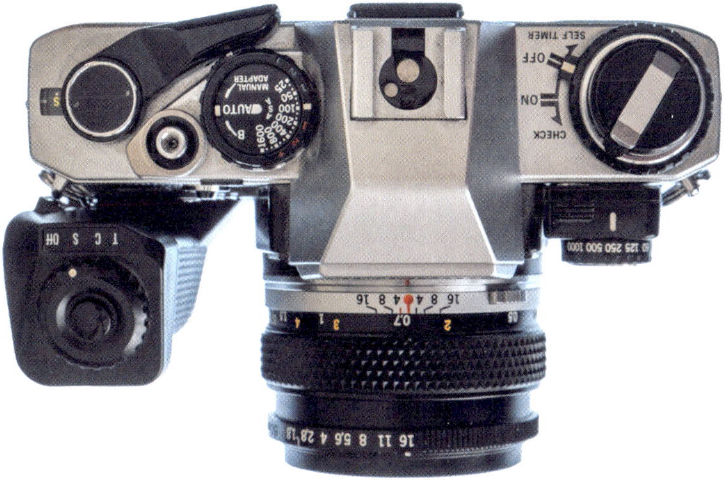

Top view of a typical SLR camera. The numbered ring on the lens nearest the body is for setting aperture. The black dial at top left sets shutter speed.

As you let in more or less light via the aperture or the shutter, the other will need to respond in the opposite direction to keep the equation balanced. For example, if you let in *more* light through your shutter, going from 1/500 down to 1/250 of a second, then you will need to let in *less* light through your aperture to compensate for that change, bringing the ratio back to the same total amount of light; the light required will be based on what the light meter suggests is the correct amount of light needed. (There are exceptions to this, and times when you may want to ignore the light meter and have more or less light in total, but

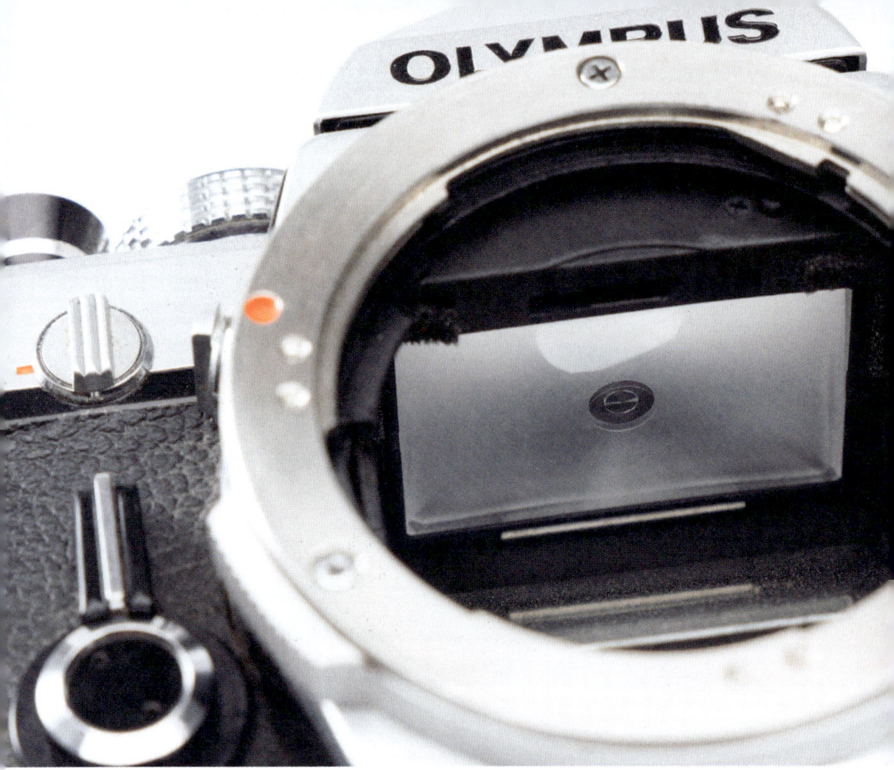

The mirror inside the body of an SLR camera reflects the light onto a prism, which in turn flips the image right-side up for the photographer to see.

generally this is the balancing act you are doing most of the time when taking photographs.)

Taking a photo with a typical SLR film camera offers a good illustration of this process. When you look through the viewfinder of an SLR, the image you see is coming from light traveling through the lens and hitting an angled mirror that directs the light up to a prism. The prism's job is to flip the image right-side up and send it out the back of the viewfinder into the photographer's eye.

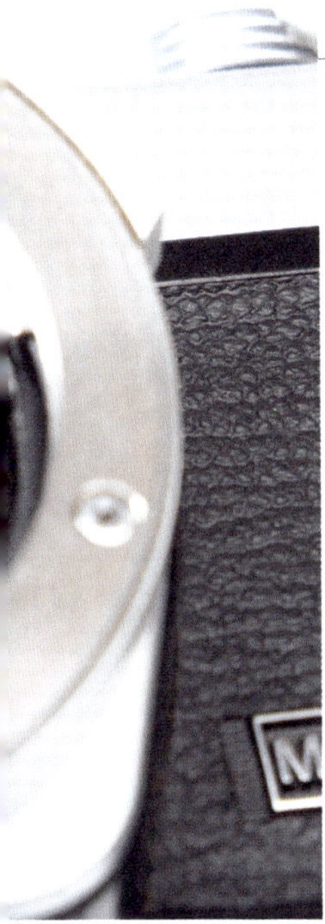

The Zeiss Ikon Super Ikonta III in this 1953 ad was a well-regarded folding camera that featured a leaf shutter. This style is not as fast as a focal-plane shutter, but it does offer superior flash synchronization.

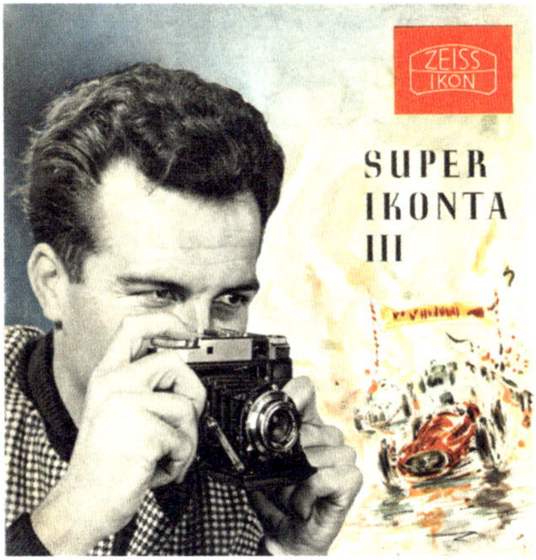

When you go to take a photograph, you will set the aperture and shutter speed settings on the camera to your desired settings. The moment you press the shutter button, the camera performs the following sequence of actions: The aperture closes to the size you selected, the mirror flips up to get out of the way of the shutter, and the shutter opens for a set amount of time to expose the film to light. Once the shutter is done, based on the set exposure time, the mirror flips back down to its resting position and the aperture opens back up to its maximum width. You then advance the film to its next unexposed frame and the camera is primed to start the sequence all over again.

CHAPTER THREE

From Bellows to Pockets

Types of Film Cameras

CAMERAS COME IN ALL shapes and sizes. Some film cameras fit inside lipstick containers for use in spy missions, while others are massive, mounted in planes and satellites for mapping the surface of the Earth. Many of the cameras used by consumers throughout photography's history were designed for one of two primary reasons: to overcome technical limitations of the period or to fill a specific niche in a style of photography.

It would take more than an entire book to present an exhaustive list of all the types of cameras ever created, but in this chapter we'll explore some of the most common categories of cameras; many have been widely available since the invention of the consumer camera. From the ghostly glow of the view camera's inverted image projected onto ground glass to the ultraportable and often funky designs of subminiature cameras, this chapter covers the categories that helped push the boundaries of what cameras could be and what they could offer us.

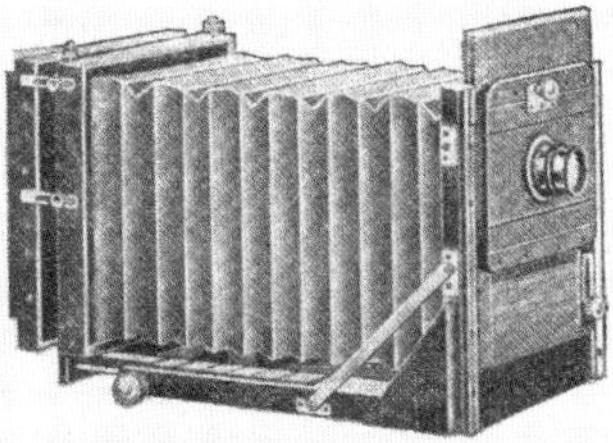

COMMERCIAL

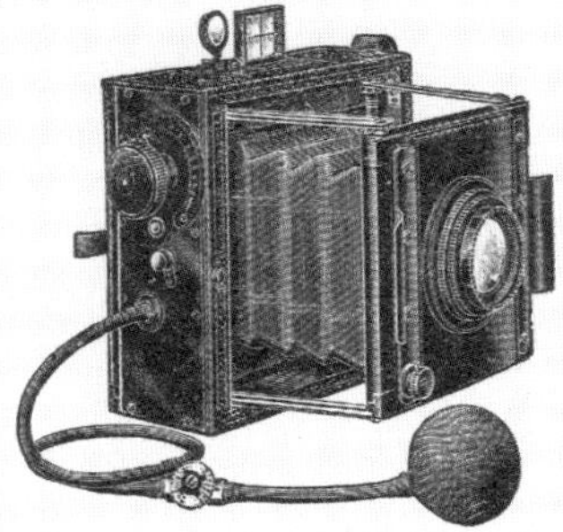

FOCAL-PLANE PRESS

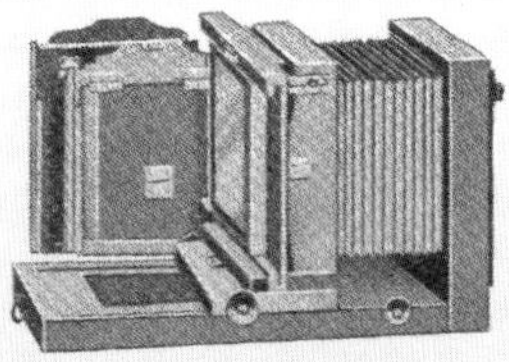

STUDIO

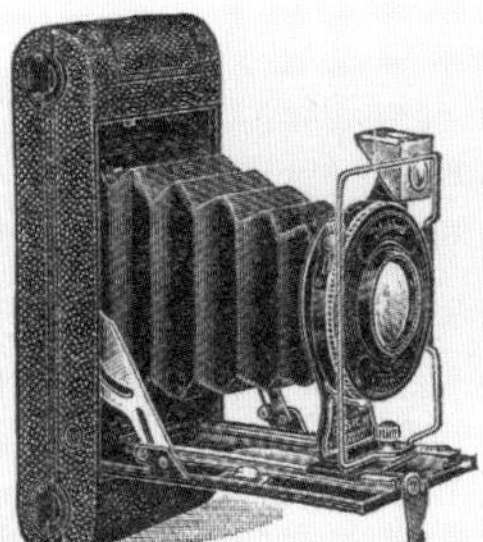

FOLDING ROLL-FILM

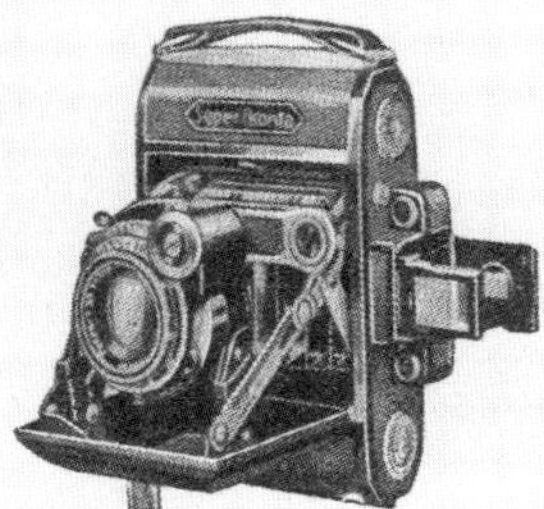

MINIATURE ROLL-FILM

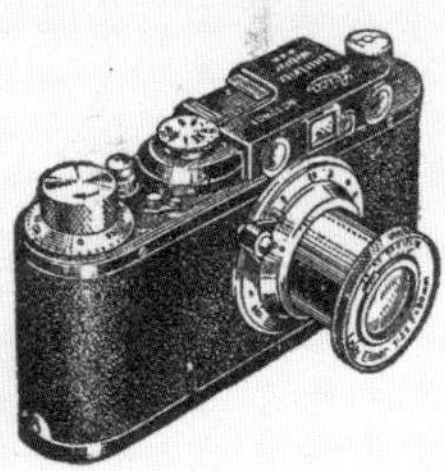

AUTOMATIC ROLL-FILM

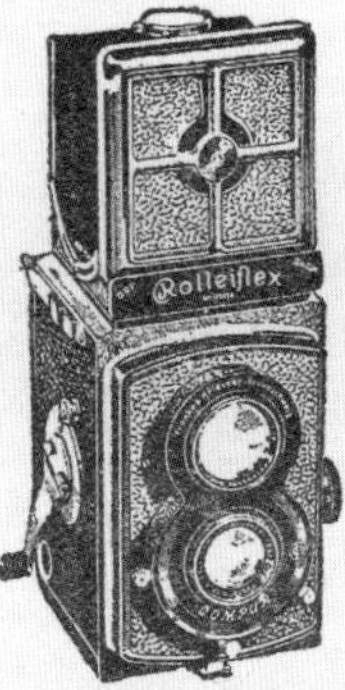

MINIATURE TWIN-LENS

Cameras have been constructed in a wide variety of shapes and sizes for shooting a number of different film formats over the years.

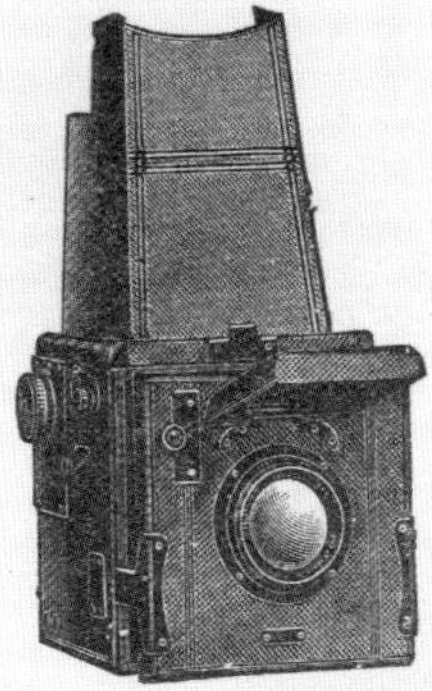

BOX-FORM REFLEX

TYPES OF CAMERAS

View Cameras

Big, clunky, heavy, and beautiful, view cameras are some of the oldest types of cameras still in use today. And for good reason: The negatives produced by these large-format cameras offer a high level of detail and the cameras themselves have visual characteristics and controls that are hard to reproduce in smaller cameras.

On one end of a view camera you'll find a lens mounted to a board; this is the front standard, which controls the lens plane. Opposite the front standard, at the other end of the camera, sits the rear standard, a pane of ground glass on which an image is projected, together with a spring-loaded mechanism that allows the operator to slide in a film holder to load the camera with film.

Connecting the front and rear standards are the bellows. Often made of fabric and reminiscent of an accordion, the bellows allow for the two ends of the camera to move independently from one another and still remain light-tight.

View cameras primarily use large sheets of film instead of rolls. The film comes in many different sizes, but the most common are 4×5 inches (10×13 centimeters) and 8×10 inches (20×25 centimeters). While film size is a characteristic that separates view cameras from other styles of cameras, the ability to move both the film plane and the lens plane independently from one another is a huge advantage view cameras have over other camera types.

The monorail and field camera are two variations on the basic view camera.

Monorail view cameras consist of a single rail system on which the front and rear standards are mounted. This allows for both standards to move in various ways independent from one another. Both the front and rear standards in a monorail view camera can rise or fall (move up and down vertically), shift (slide horizontally), tilt, and swing. This allows

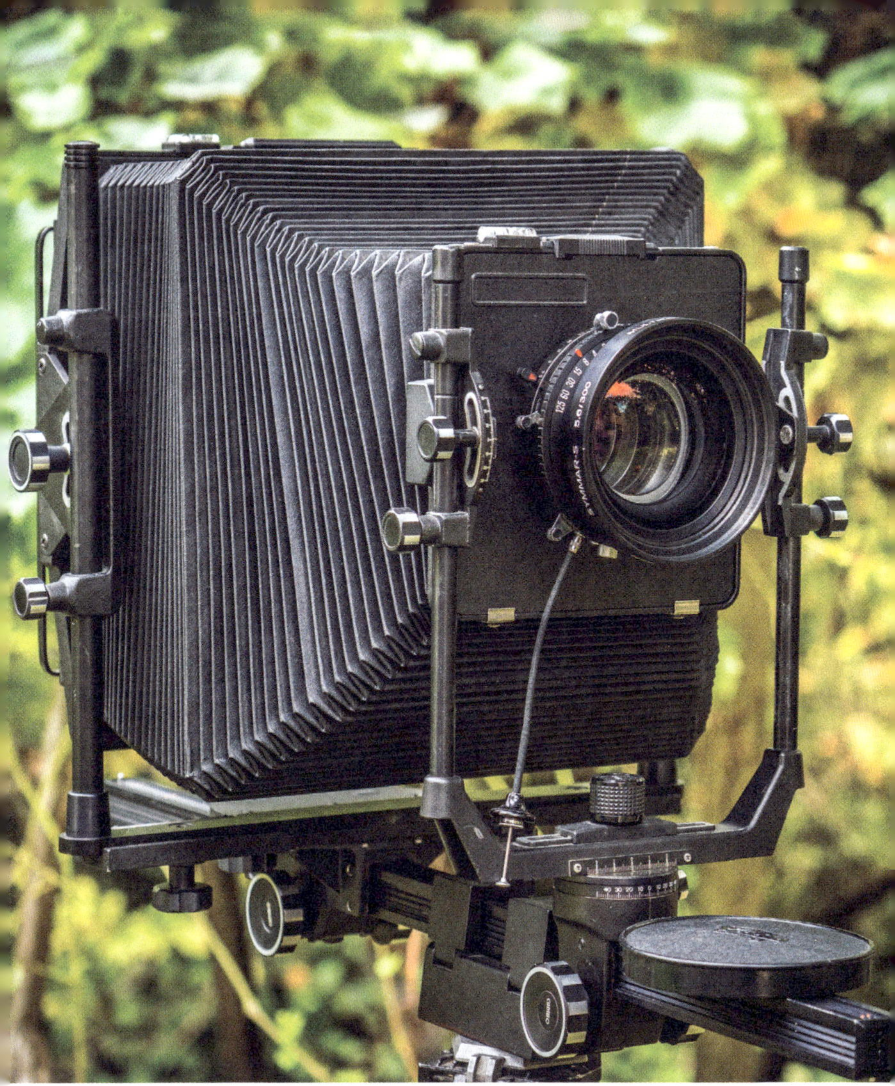

While larger and bulkier than their field camera counterparts, monorail cameras allow for a much larger range of front and rear standard movement and corrections to be made compared to field cameras.

for the unique ability to correct for perspective and control the focus plane. These cameras tend to be bigger, bulkier, and more cumbersome to handle, which can be a disadvantage for portability.

In contrast to monorail view cameras, field cameras tend to be more compact, consisting of a rail system mounted to a hinged platform that

Unlike monorail cameras, field cameras collapse into a self-contained box, allowing for much easier storage and transportation.

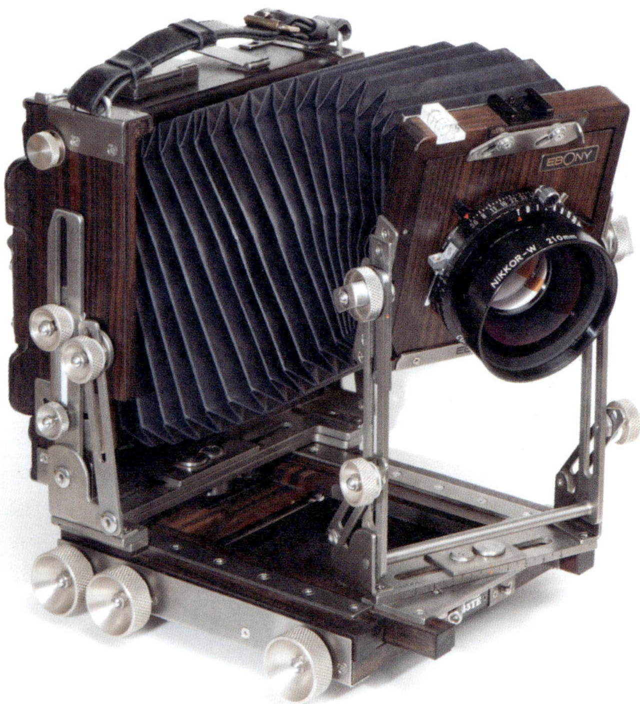

can fold up neatly into a compact, self-contained box when not in use. While more portable (even though still relatively large and clunky), field cameras come with the tradeoff of having only the front standard free to move while the rear standard is locked in place. These cameras were often used by the press before portable 35mm cameras became available.

Reflex Cameras

When you think of film cameras, you likely imagine a reflex camera. With their (relatively) small and portable form, interchangeable lenses, and diverse accessories, reflex cameras became the cameras of choice for everyone from fashion photographers to journalists, to hobbyists in the twentieth and early twenty-first centuries.

The concept behind film reflex cameras is fairly simple: Light enters a lens and is redirected into a viewfinder via an angled mirror. As mentioned in chapter 1, mirrors used to redirect an image into a camera have been employed since the early days of the camera obscura, and the concept persisted down to the smaller mechanical cameras of the twentieth century. Reflex cameras fall into two basic categories: twin-lens reflex cameras (TLRs) and single-lens reflex cameras (SLRs). Both have unique properties, but SLRs became far more commercially successful than the twin-lens models.

SLR

In an SLR, the lens, mirror, shutter, and film are all aligned perfectly behind one another in a single path through the camera. As the light enters the lens, an angled mirror reflects it up into a pentaprism or sometimes onto a piece of ground glass (which sits above the elements listed). In the case of cameras with a pentaprism, the light directed

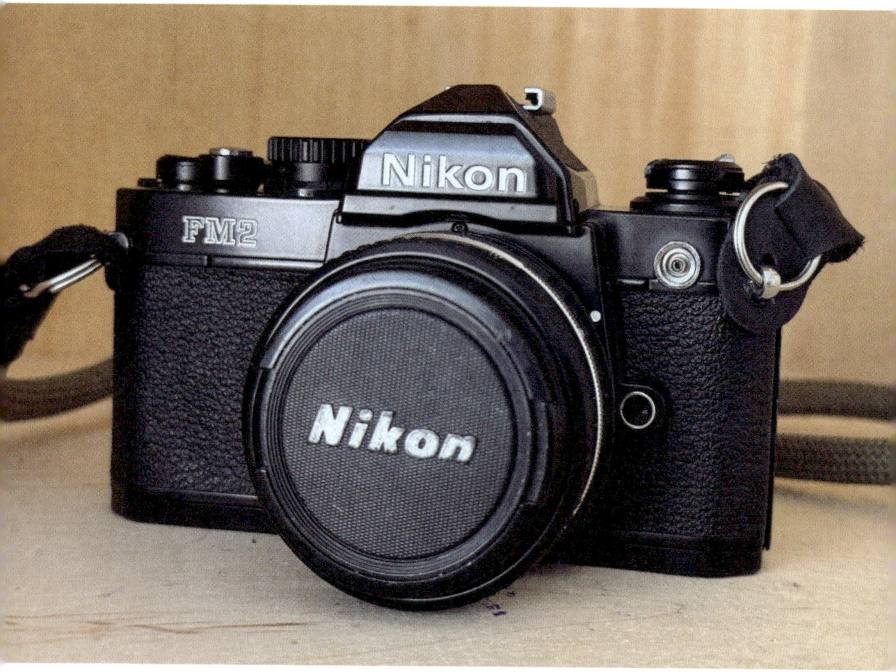

The Nikon FM2n is a fully mechanical SLR known for its blistering-fast top shutter speed of 1/4000 and for being an incredibly reliable alternative to Nikon's flagship F-series cameras.

upward by the mirror enters the prism and is flipped and then directed out the back of the viewfinder. This allows the photographer to see the image right-side up rather than upside down. If the camera does not have a pentaprism, like many medium-format cameras with waist-level finders, the image is simply reflected by the mirror up to a piece of ground glass and viewed backward.

When the shutter button is pressed to take a photo, the mirror flips up to get out of the way, the shutter opens, and the light travels straight through the camera onto the piece of film in the back. The classic click we associate with taking photos is most often the mirror flipping up to get out of the way and then slapping down in place after the film is exposed.

SLR cameras allow the photographer to see exactly what the lens is seeing, rather than the approximation of TLRs or rangefinders, where the viewfinder is separate from the lens system. The downside of SLR is that the whole system itself is more mechanically complex, with far more moving parts, and when an image is taken there's a brief blackout while the mirror has flipped up.

Despite these limitations, the SLR design has won out over the TLR. This is likely due to the advantage of seeing through the lens what will produce the final photograph.

SLRs come in a number of film sizes and formats, but the most common are medium format and 35mm.

TLR

Twin-lens reflex cameras have two lenses instead of one, with the lenses stacked vertically, one on top of the other. The topmost lens is paired with a mirror behind it, and the mirror and lens work in tandem to project an image onto a piece of ground glass that sits atop the camera. Typically, the photographer stands with the camera at waist level and looks down onto the ground glass to compose and focus the image.

Most TLRs are far less mechanically complex than SLRs. Most TLRs have leaf shutters built into the second lens, which means that these cameras don't have to execute a coordinated series of mirrors flipping up, shutters opening, and apertures closing down. Instead, once the shutter button is pressed, the TLR trips the leaf shutter in the lens, acting as both the aperture and the shutter and exposing the film to light. This is far less complex than the SLR, with reduced potential for

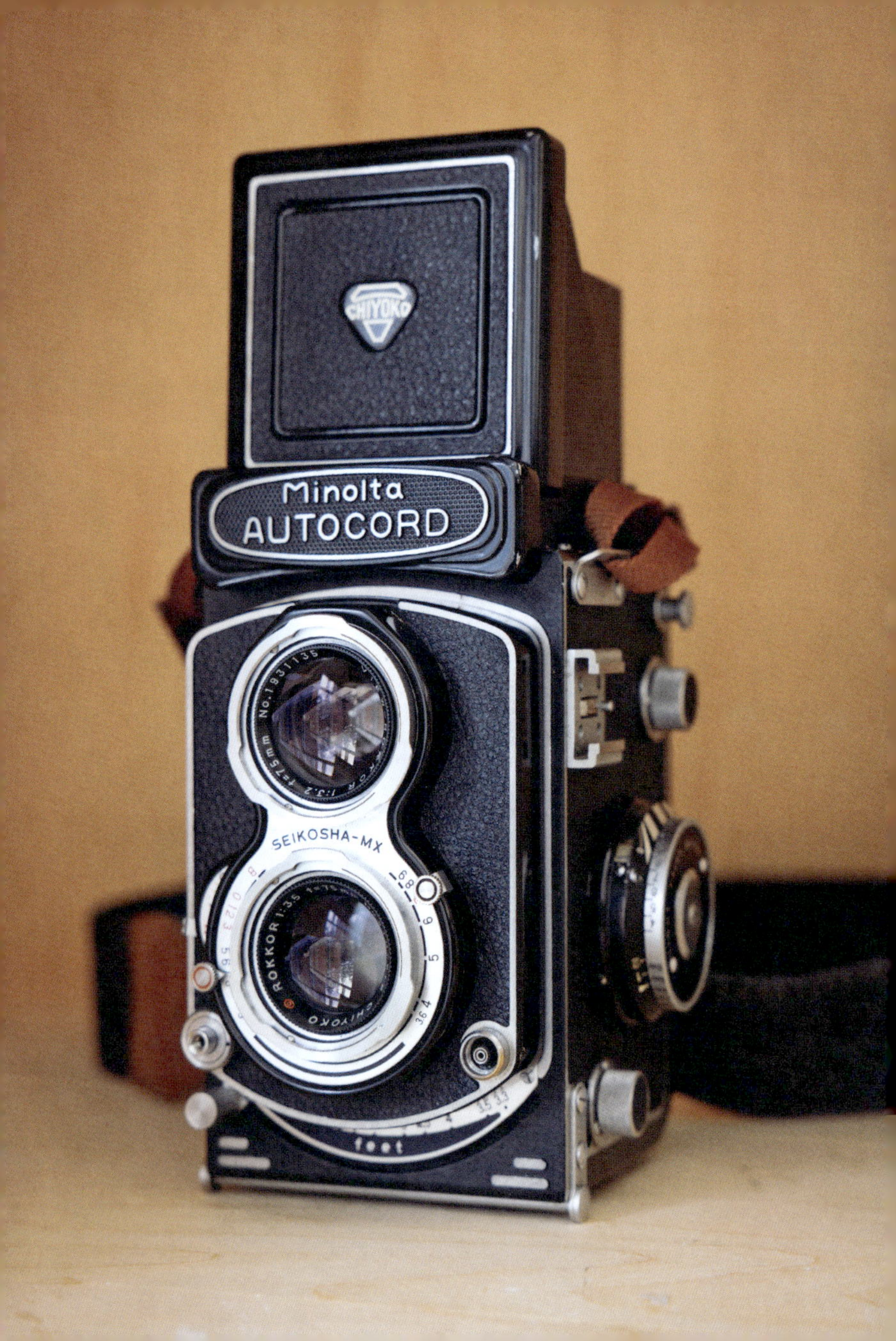
CHIYOKO
Minolta
AUTOCORD
SEIKOSHA-MX
ROKKOR 1:3.5
CHIYOKO
feet

breaking or wearing out individual parts. And because there's no mirror to move out of the way, you can see the image throughout the entire exposure process—a huge advantage, particularly during long exposures.

TLRs aren't perfect, though, and one particularly glaring design compromise is that the camera suffers from parallax issues: Using two lenses means that the image you see in the viewfinder is not the image the camera records. This problem is more apparent the closer you physically get to the object you're trying to photograph.

Leaf shutters and focal-plane shutters also have their own advantages and disadvantages. While leaf shutters can sync with flashes at any shutter speed, they aren't as fast as focal-plane shutters—usually topping out at only around 1/500 of a second—and the focal length of lenses usable with TLRs is limited compared to their SLR counterparts.

Despite these limitations, TLRs were a popular choice among photographers, offering a portable, flexible alternative to view cameras before the SLR was widely available. TLR cameras typically are medium format, though some 35mm TLR cameras do exist.

Produced from 1955 to 1966, the Minolta Autocord was a fully mechanical twin-lens reflex camera designed to compete with high-end TLRs such as the Rolleiflex. The Autocord is known for its high-quality optics and reliable construction.

Rangefinders

If you were a photographer in the first half of the twentieth century and wanted a camera that was light, portable, and quick, had accurate focusing, and produced high-quality images, you likely chose to shoot with a rangefinder camera. These cameras get their name from the rangefinder included in their design. This device allows the photographer to measure the distance to the subject in order to achieve precise focus.

The most common types of rangefinders display two images when looking through them: One is the image of the outside world and the other is a ghostly, transparent copy overlaid on top. As you turn the rangefinder gears, the two images move closer or farther apart from one another. Once the two images overlap perfectly, you look down at the gear of the rangefinder to find the distance given. If the rangefinder is a decoupled type, the photographer takes the distance given and sets that distance on the lens. Later models have coupled rangefinders, meaning the focusing ring of the lens is mechanically connected to the rangefinder, so all you need to do is turn the focus of the lens to get the images to overlap, giving you a perfectly focused image.

SLRs and TLRs display an image in the viewfinder that comes through a lens and then is reflected into the viewfinder via a mirror. In contrast, a rangefinder's viewfinder is a viewport near the lens that, in most cases, goes straight through the camera body. Because of the lack of space needed for a mirror box and pentaprism, rangefinders tend to be smaller, lighter, and much quieter than reflex cameras.

While these characteristics make them a desirable platform to shoot with even today, rangefinders do have a few limitations that made them lose ground to SLRs in the second half of the twentieth century. For example, they tend to be limited to fixed-focal-length (prime) lenses because their design doesn't allow for easy adaptation to zoom lenses. And because the image you see through the viewfinder isn't coming

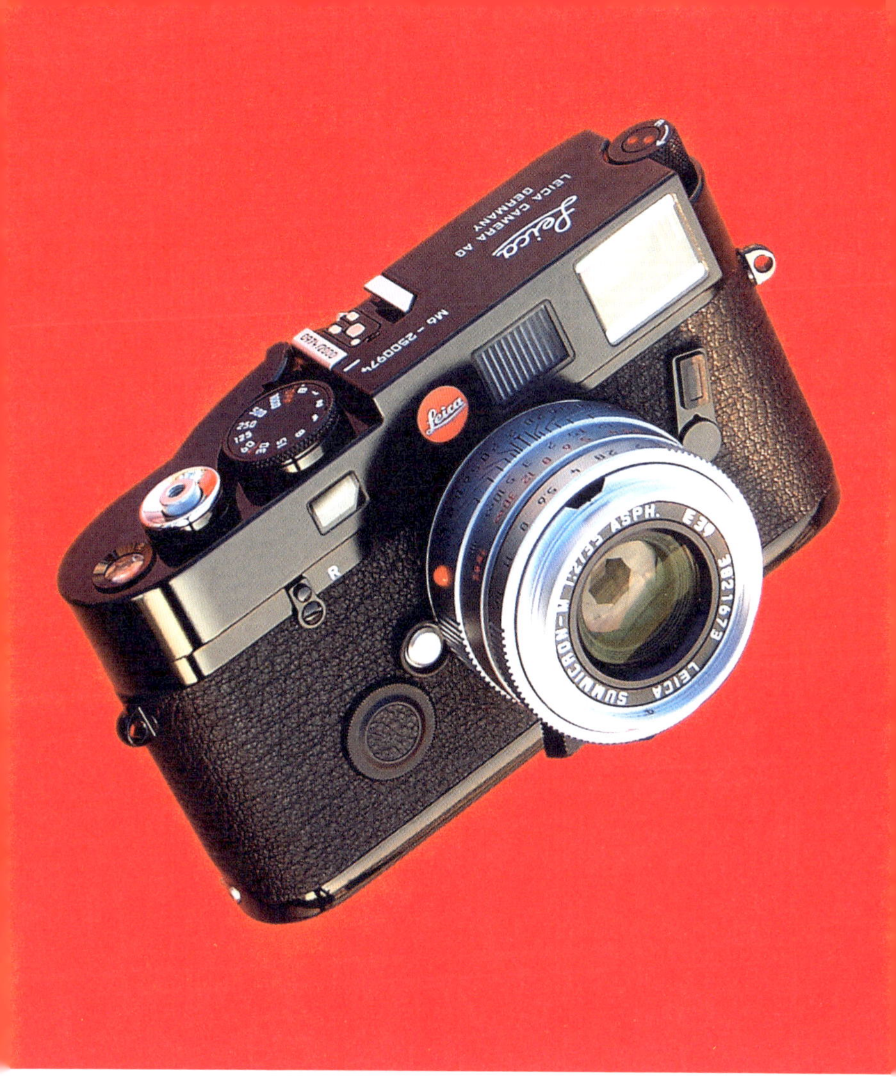

With their large, bright viewfinders and precision engineering and manufacturing, Leica set the standard for what small, portable rangefinder cameras could be.

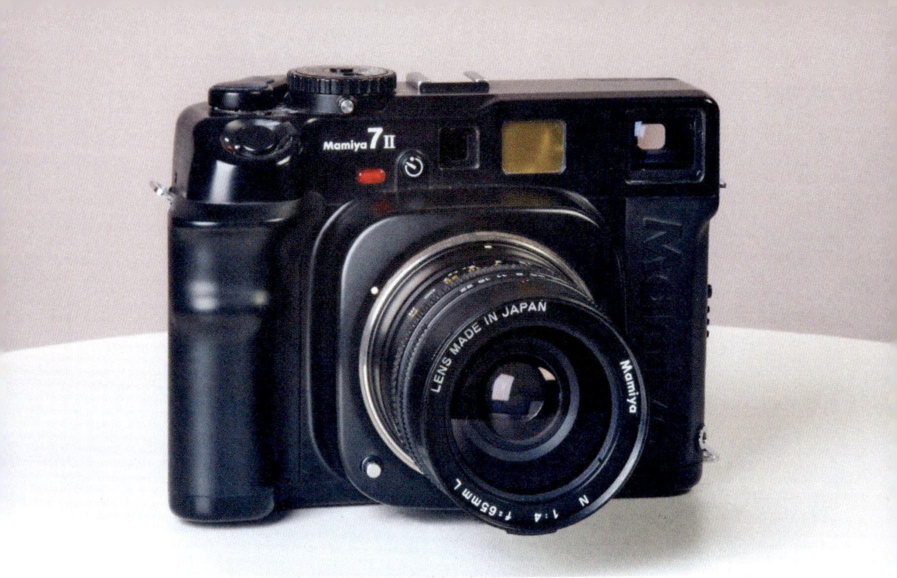

Not all rangefinders are limited to 35mm film cameras. The Mamiya 6 and its bigger sibling, the Mamiya 7, are examples of compact medium-format rangefinder cameras.

through the lens, they are susceptible to parallax, which makes them harder to shoot with when the subject is close to the camera.

Many camera manufacturers produced a number of different types of rangefinder cameras, such as Kodak's No. 3A Autographic Special, the world's first commercially marketed rangefinder, released in 1916; and Argus's C3 rangefinder (lovingly nicknamed "the Brick"), which was the most successful rangefinder in America, selling millions of units. Of all the rangefinder cameras produced, however, Leica and their M series have been synonymous with this style. Not only did Leica set the standard for quality and design—later copied by Japanese brands like Nikon and Canon—but the standard for quality optics, manufacturing, and engineering endures, with film Leicas still demanding high price tags in the used camera market.

Folding Cameras

Popular from the latter part of the nineteenth century through the 1940s, folding cameras were an elegant solution to the technical limitations of the lens and camera technology of early camera designs.

To correctly and accurately project a focused image onto the surface of a piece of film, lenses require a specific distance between the film and the lens. Early lens designs, coupled with larger negative sizes, required a much greater distance, calling for a larger camera.

Folding cameras are capable of folding down into smaller, sturdy form factors for transportation and storage between uses. They typically consist of a lens connected to the camera via a set of bellows. That lens is also attached to a hinged platform, allowing the lens-and-bellows system to slide out and extend from the camera body when opened and

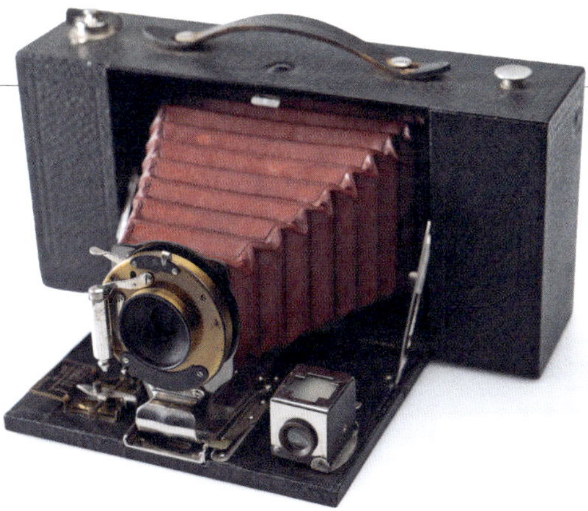

Produced from 1909 to 1915, the Kodak No. 3A Folding Brownie is a visually striking example of a folding camera, with its red leather bellows and black leather body. It shot 122 film and listed for a retail price of $10.

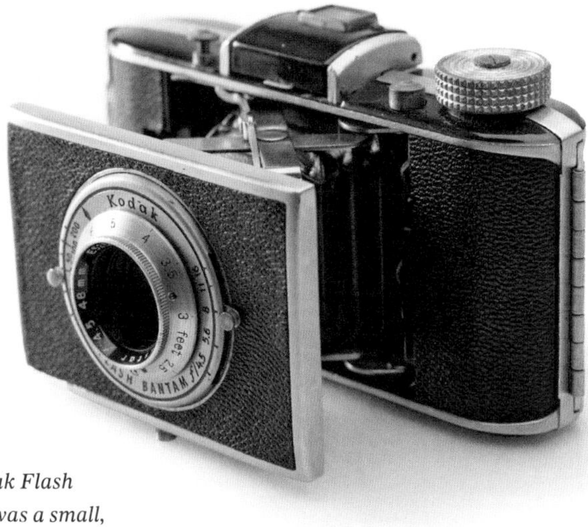

The Kodak Flash Bantam was a small, stout pocket camera that shot on 828 film, a roll film that was Kodak's attempt to produce an alternative to 35mm, but the film format never quite took off.

retract into the camera body, compactly folding up when closed.

The distance required between lens and film shrank as lens and camera technology improved, along with the gradual rise of 35mm as the dominant film choice. Cameras could be smaller and more compact, reducing the need for portability and functionality offered by folding cameras. While the design became less necessary, it never went completely away: Polaroid used folding camera designs for many of their cameras well into the 1980s and 1990s.

This folding camera design was applied to all types of cameras, from view cameras like the famous Graflex Speed Graphic to rangefinder cameras like the Kodak's No. 3A Autographic Special and the Kodak Retina, as well as instant cameras like the classic Polaroid SX-70.

Instant Cameras

For those who grew up in a world where digital cameras have always existed, it can be hard to imagine a time when photographs weren't always immediately viewable after being taken. Unlike digital photography, film photography relies on a complex process of chemical reactions to turn film into an image; often that process involves waiting for the entire roll of film to be shot, then sending it off, and waiting for a professional to develop the film. Instant cameras solve both of these problems.

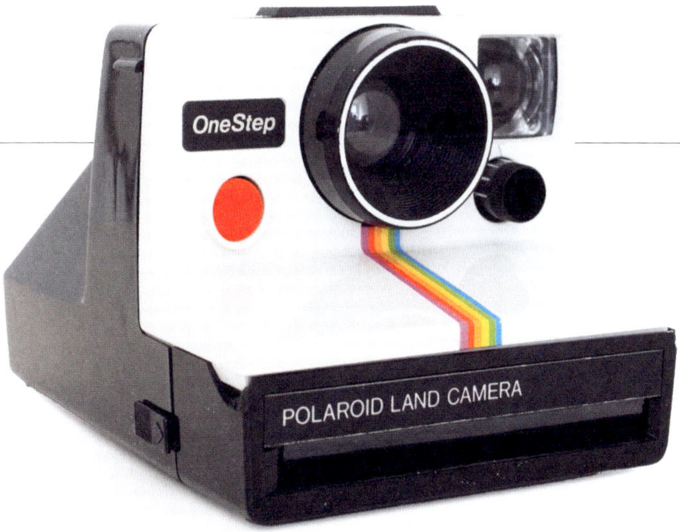

The Polaroid OneStep SX-70 Land Camera, with its distinctive white body and rainbow stripe, has become an icon that transcends the world of photography. For example, it was the inspiration for Instagram's first logo.

Developed by Edwin Land in the 1940s, instant cameras take one image at a time on a specially designed film that contains all the chemicals needed to develop the film. The design removed the need to get images developed at a lab or back home in a darkroom: You could take a photograph and within minutes or even seconds have a positive image develop on the spot. Land went on to found Polaroid in the 1940s. The company produced instant cameras and the film needed to use them. The first was the model 95 Polaroid Land camera, which sold 1.5 million units and is widely considered the first commercially successful instant camera.

Other manufacturers eventually jumped into the instant camera market. Some, such as Konica and Minolta, created instant cameras that used Polaroid film, while others, such as Kodak and Fuji, created their own types of film to use in their cameras.

In 1972 Polaroid released the SX-70 camera alongside a new third-generation Polaroid film. The SX-70 was a foldable SLR instant camera that, despite its high price tag of $180 (equivalent to $1,400 in 2025), would become one of the most iconic instant cameras ever made. The camera went through several models over the years, all sharing the same basic design, but each introduced more advanced features. In 1978, the updated SX-70 featured the Sonar OneStep sonar-based autofocus system, a new design that made it the first commercially available SLR to include autofocus.

Fuji jumped into the instant film market in the 1980s and eventually launched their Instax cameras and film. Based on the technology Polaroid introduced in their third-generation film, Instax photos looked similar to Polaroids but offered a different aspect ratio taken with the company's new, smaller cameras. Fuji rose out of the collapse of the analog camera and film market of the early 2000s slightly less wounded than Kodak and Polaroid, enjoying a resurgence of interest in the instant camera market. They've continued to release new instant cameras into the twenty-first century.

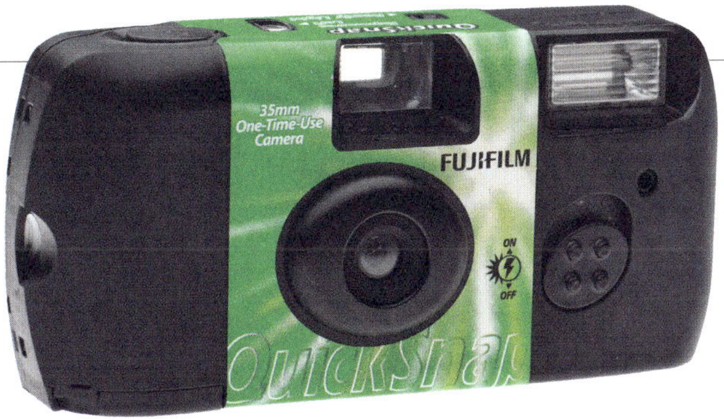

The Fuji QuickSnap disposable camera opened new possibilities for low-stakes image making in a variety of new situations and is still available today.

Point-and-Shoot Cameras

Photography has always been a technical medium. The photographer has to pay attention to any number of settings and make decisions before snapping a photograph. What is the lighting like? What aperture or shutter speed do I need? What ISO (film sensitivity) is required? Is the subject in focus? What composition do I want? The list can seem endless. For hobbyists, professionals, and serious amateurs alike, controlling these settings and decisions may be important, but the vast majority of people just want to press a button and capture a moment. That's where point-and-shoot cameras come in.

Kodak understood early on that photography needed to be simple if the general consumer was going to adopt it. In 1888 and 1900, respectively, the company released the Kodak No. 1 and Brownie cameras. Both had fixed apertures and fixed shutter speeds; the No. 1 came preloaded with film. All you had to do was press the button and let Kodak do the rest (which also happened to be their slogan).

Point-and-shoot cameras came into their own from the 1980s through the early 2000s, making up the bulk of the cameras sold during that

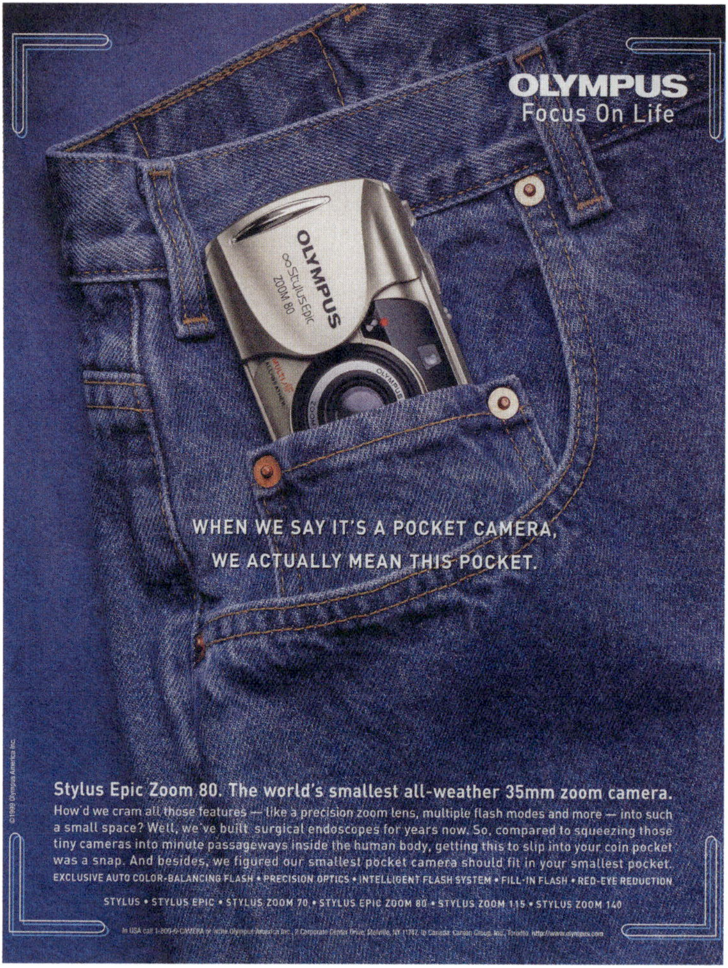

Known for its small, compact form factor, accurate autofocus, and weather sealing in the Epic line, the Olympus Stylus cameras wound up being a successful line of cameras for the company.

period. They were small and pocketable, with simple functionality that removed all the work for the operator; many even had fixed apertures and automatic shutters. They either had fixed focuses or came with autofocus, and many automatically loaded, advanced, and rewound the film for you.

In 1978 Canon released the Sure Shot/AF35M, the first "lens-shutter" 35mm camera with active autofocus (eighteen months after Konica released their passive autofocus system). The innovative focus mechanism consisted of a diode that emitted near-infrared light that was "seen" by a sensor on the camera, which in turn measured the distance to the subject and drove the focus of the lens. The system became incredibly popular and kicked off Canon's widely successful Sure Shot line. A little over a decade later, Olympus became a household name in the point-and-shoot market with their Stylus line, selling an astounding 20 million units in the first ten years of production.

The point-and-shoot market didn't always mean you had to invest in a camera that would last forever. In 1986 Fuji introduced the first single-use camera (a disposable camera) with their Fujifilm QuickSnap. The camera had a plastic body with a cardboard exterior. The shutter speed was set at 1/140 of a second, the lens had a fixed aperture and did not need to be focused, and the camera came with a built-in flash. It was preloaded with film and sealed shut. Once you were done shooting the roll of film, you dropped the entire camera off at a photo lab that removed the film and disposed of the camera body. The concept was so successful that many of the major film producers of the time (including Fuji, Agfa, and Konica) went on to produce a variety of disposable cameras: panoramic, waterproof, black and white. It was the perfect item to take on vacation, hand to children to play with, or set out on tables at weddings.

The point-and-shoot camera brought the masses to photography, lowering the barrier to entry and removing the need to understand how a camera works to enjoy the act of capturing moments. That spirit lived on long after film fell out of popularity, with point-and-shoot digital cameras in the early 2000s and cellphone cameras of today.

Subminiature Cameras

Supersmall, portable, and reminiscent of old spy movies, subminiature cameras include any cameras that shoot on film smaller than 35mm. These cameras make ultraportability the top design priority.

Subminiature cameras come in all shapes and sizes and use a wide variety of film sizes, but any history of this type will begin with the Minox series. Designed by Latvian engineer Walter Zapp in 1936, Minox cameras were a small, compact line said to be the most widely used spy cameras on the planet. The Minox B, for instance, measuring a mere 4×1.1×0.6 inches (10×2.8×1.6 centimeters), shot film one-quarter the size of 35mm and easily fit in the palm of your hand. Its close-focusing lens made it excellent for photographing documents, a handy trick for those involved in espionage.

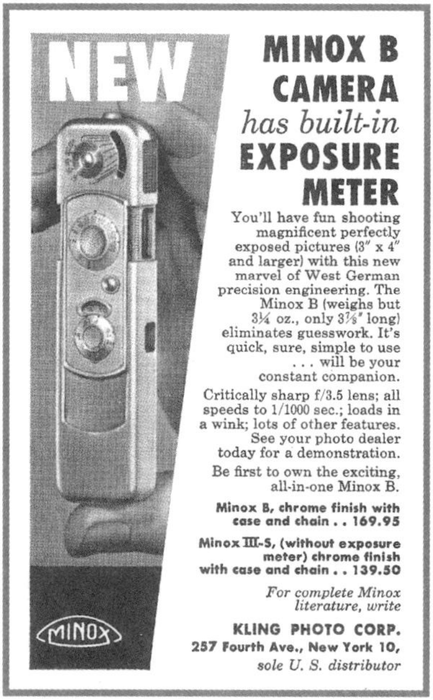

Known as the classic "spy camera," the Minox wasn't limited to espionage work: It was also marketed as the perfect portable, everyday carry camera.

Not all subminiature cameras were relegated to spy work, however. First developed in the 1930s, HIT-type cameras began to take off in popularity in Japan in the wake of World War II in response to a shortage of raw materials and a lack of

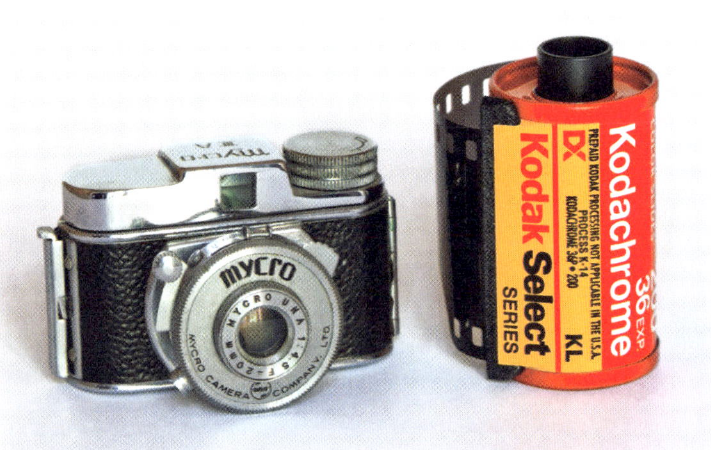

The Mycro camera is a classic example of the cute subminiature cameras produced by Japan in the 1940s.

disposable income. HIT-type cameras are, frankly, adorable. Featuring chrome parts wrapped in leatherette material, they look like someone shrank a classic rangefinder down in size. Most of these types have a fixed-aperture lens with no focus controls and a single-speed shutter (though some do have a bulb mode); they shoot 17.5mm (half-sized 35mm) rolls of film. The most famous of these types of cameras was Tougodo's HIT, which is where this style of camera gets its name.

In 1972 Kodak introduced a smaller, more compact alternative to 35mm film they labeled 110. Kodak's 110 film format consisted of a 16mm film strip packaged in a cartridge. The design was essentially a miniature version of their previously manufactured 127 film system. The 110 film was launched with the debut of Kodak's new Pocket Instamatic. While the format enjoyed some commercial success—and other camera makers like Minolta jumped in and used the format in cameras like the Minolta 110 SLR—it was never widely adopted.

Befitting all the quirkiness of the 1980s, in 1982 Kodak introduced the completely novel Kodak Disc cameras. Small and sleek, with a profile similar to a Sony MiniDisc player, Disc cameras were meant to usher in a new era of amateur photography for the masses. They got off to a hot start, with over 8 million cameras sold in the first year after their launch. Film came in a thin plastic container that held a circular disc with fifteen tiny negative frames circling its outside edge. Although the cameras were packed with cool technology and loads of innovative features, the platform was plagued with problems: The film required an investment in all-new development and printing machines, and most film labs weren't willing to take on the added cost and learning curve required to develop the film. The results were laborious, hand-developed negatives printed on machines that didn't have the correct number of lenses and, as a result, produced poor prints. While 25 million Disc cameras were sold, after just six years of production Kodak pulled the plug.

With styling reminiscent of the Sony MiniDisc players, Kodak Disc cameras were an impressive feat of engineering, but ultimately their film format was a dead end for Kodak.

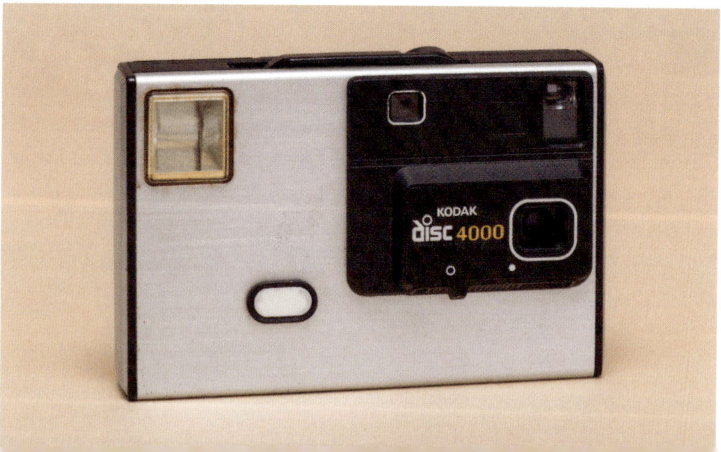

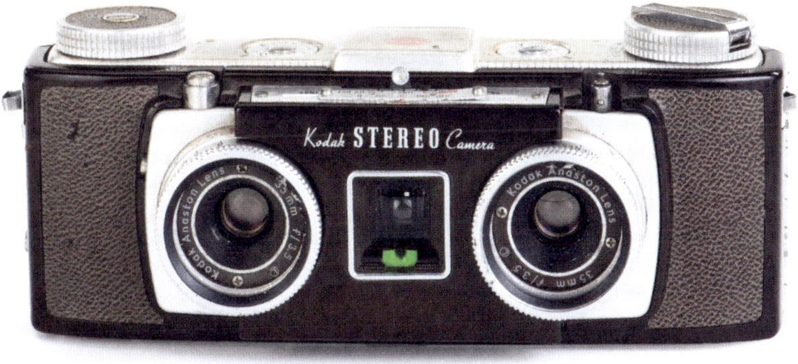

With a list price of almost double that of Kodak's camera didn't stop Stereo Realist from being a successful product, often sported by celebrities in the company's magazine ads.

Stereoscopic Cameras

Long before virtual reality headsets and 3D movies, stereoscopic cameras created photographs that seemed to emerge from the print surface as the viewer looked at them. Humans perceive the world in three dimensions through binocular vision: With two eyes that see the world at ever-so-slightly different angles, our brains use those differences to perceive depth.

Sir Charles Wheatstone first described the concept of stereoscopy in 1832. Six years later he made the first stereoscope, a device that allowed viewers to look at two images side by side and experience them in three dimensions. Wheatstone's stereograph was initially used to look at line drawings. It was large and clunky and used mirrors to create the stereoscopic effect. Sir David Brewster improved on the stereoscope design with a lenticular stereoscope in 1849, replacing the mirrors with

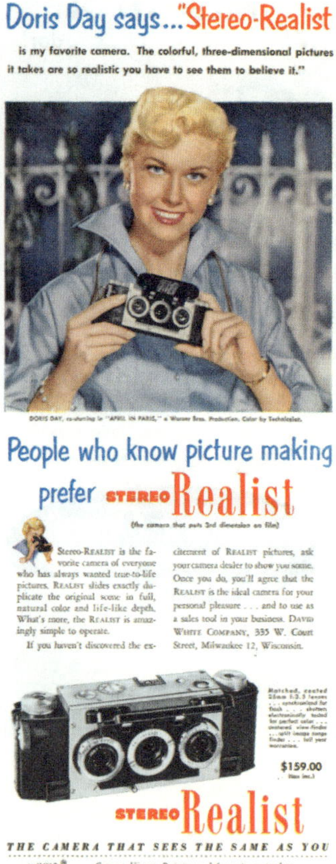

lenses that the viewer looked through, making it more compact and practical for wider adoption.

Stereoscopic photography became popular in the 1850s, and commercial stereoscopic cameras were widely available by the end of the nineteenth and beginning of the twentieth century. An entire industry popped up around taking and selling stereoscopic photographs, which were often mounted on cardboard and sold to consumers. Not only could you now see images of faraway places, you could "experience" them in 3D! With cameras like the Kodak Brownie Hawkeye Stereo and the Verascope, making stereo images was easy and affordable for consumers. Stereo cameras operated similarly to their nonstereo counterparts, but they differed in that they consisted of two lenses mounted next to one another, taking two images side by side. Stereo cameras became

With a list price of $84.50, Kodak's Stereo camera was a relatively successful camera line, selling roughly 100,000 units during its production run.

so popular that, by the onset of World War I, it's estimated that half of all amateur cameras were stereoscopic.

Stereo cameras lost ground as the century progressed, but they found new life in the 1950s with the release of the Stereo Realist camera. Developed by Steton Rochwite and produced by the David White Company of Milwaukee, the Stereo Realist was a 35mm rangefinder camera that sold so well it spurred a resurgence in stereo camera development from a number of major brands during the decade. Interest in dedicated stereo cameras didn't last, though, and in the second half of the twentieth century the cameras were relegated to niche markets. They never fully disappeared, even when cameras transitioned to digital in the 2000s.

Panoramic Cameras

From large sweeping landscapes to the hustle and bustle of busy cityscapes to large group photos, the desire to shoot wide has been around almost since photography was introduced. While the photographer can take multiple images of a scene to try to cover a desired field of view, panoramic cameras were developed to allow the photographer to capture a wide field of view in a single image. While there is no tightly defined set of requirements to describe a panoramic image, it is generally agreed that a panorama is an image that either captures a field of view wider than the human eye can see or has an aspect ratio where the width is at least two times that of the height.

The first commercially available panoramic camera was the Al Vista released in 1897 by the Multiscope & Film Corporation of Burlington, Wisconsin. The camera incorporated a swing lens connected to soft, flexible leather bellows and was driven by a spring that swept the lens across the film in a 180-degree arc. The shutter speed was determined by how fast the sweeping action occurred, and the speed of the sweeping action was controlled by selecting from a set of different-sized fans that created aerodynamic drag. The Al Vista came in various sizes and models and was successful enough to catch the attention of Kodak, which designed and released their own line of swing-lens panoramic cameras just three years later, with the debut of the Kodak Panoram line.

The Cirkut camera offered a variation for photographers who wanted to shoot on large-format film. Whereas swing-lens cameras keep the camera body and film stationary during the exposure while the lens moves, the Cirkut employs a full-rotation design: The lens in these cameras is fixed but connects the film advance to the camera base. As the camera body rotates to capture a scene, the film is pulled through the camera at a rate that matches the camera movement. Originally manufactured by Rochester Panoramic Camera Company (which was eventually purchased by a subsidiary of Kodak), the Cirkut came in several models shooting different lengths of negatives, the largest up to 20 feet (6 meters) long!

As 35mm film came to dominate the second half of the twentieth century, small, portable 35mm panoramic cameras continued to develop. Models like the Widelux and Horizon, introduced in the 1950s and 1960s, were produced until the early 2000s, featuring swing-lens designs and sizes comparable to a typical SLR camera of the period, but offering images three times as long as a typical photograph (twelve exposures on a thirty-six-exposure, 35mm roll of film).

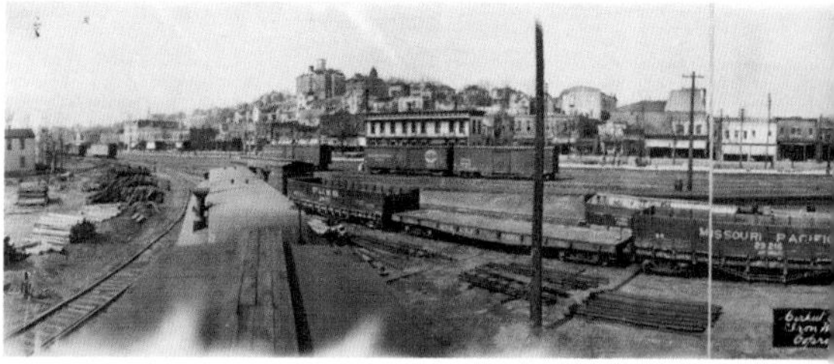

This view of a railyard was created with a Cirkut camera, the design of which used a rotating body to capture photos up to 20 feet (6 meters) long.

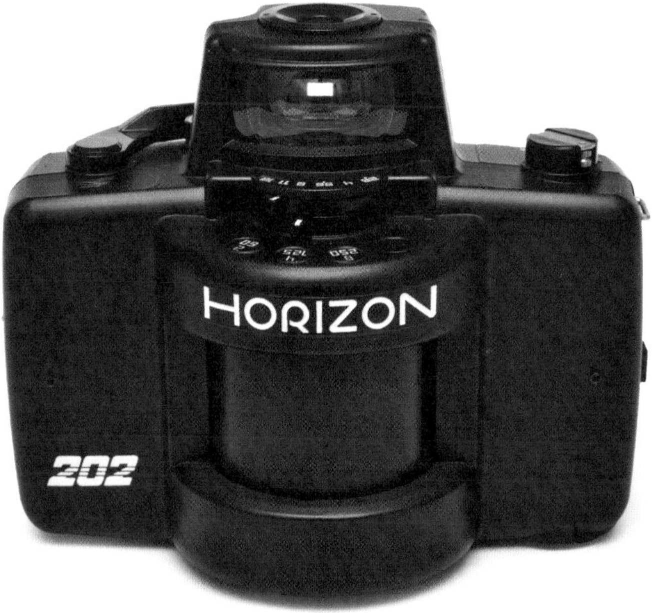

The Horizon was a Russian panoramic camera of the swing-lens variety that shot a 120-degree photograph on 35mm film.

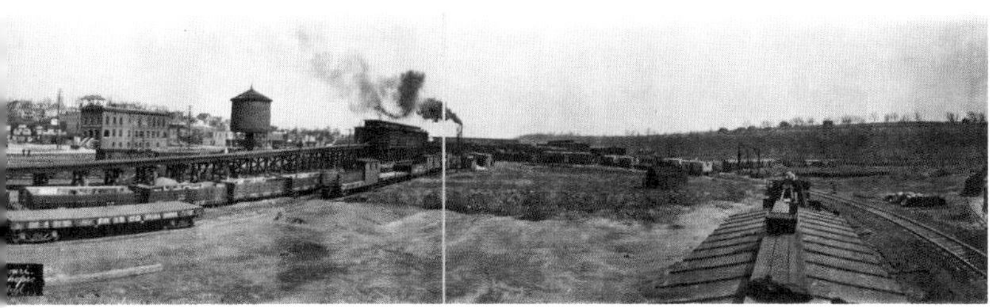

CHAPTER FOUR

Industry Icons

Notable Brands

THE INVENTION OF PHOTOGRAPHY was shaped by scientists, artists, and engineers looking for ways to permanently record images of the world around them. Once the medium was invented, however, the commercialization of photography was shaped more by brands competing for sales than by individuals. Kodak arguably paved the way with their affordable cameras well before other brands. Not all companies emerged like Kodak, that is, solely for the purpose of creating photographic equipment—often companies developing new technology, like manufacturing optical equipment or creating commercial chemicals, were naturally positioned to apply their manufacturing and research knowledge to the emerging camera market.

The twentieth century saw the rapid expansion of the commercialization of film photography, with an explosion of camera, film, paper, chemical, and lens brands innovating and developing new products to

While Agfa's longest-lasting product is its Rodinal developer, the company also offered a number of quality film stocks and cameras.

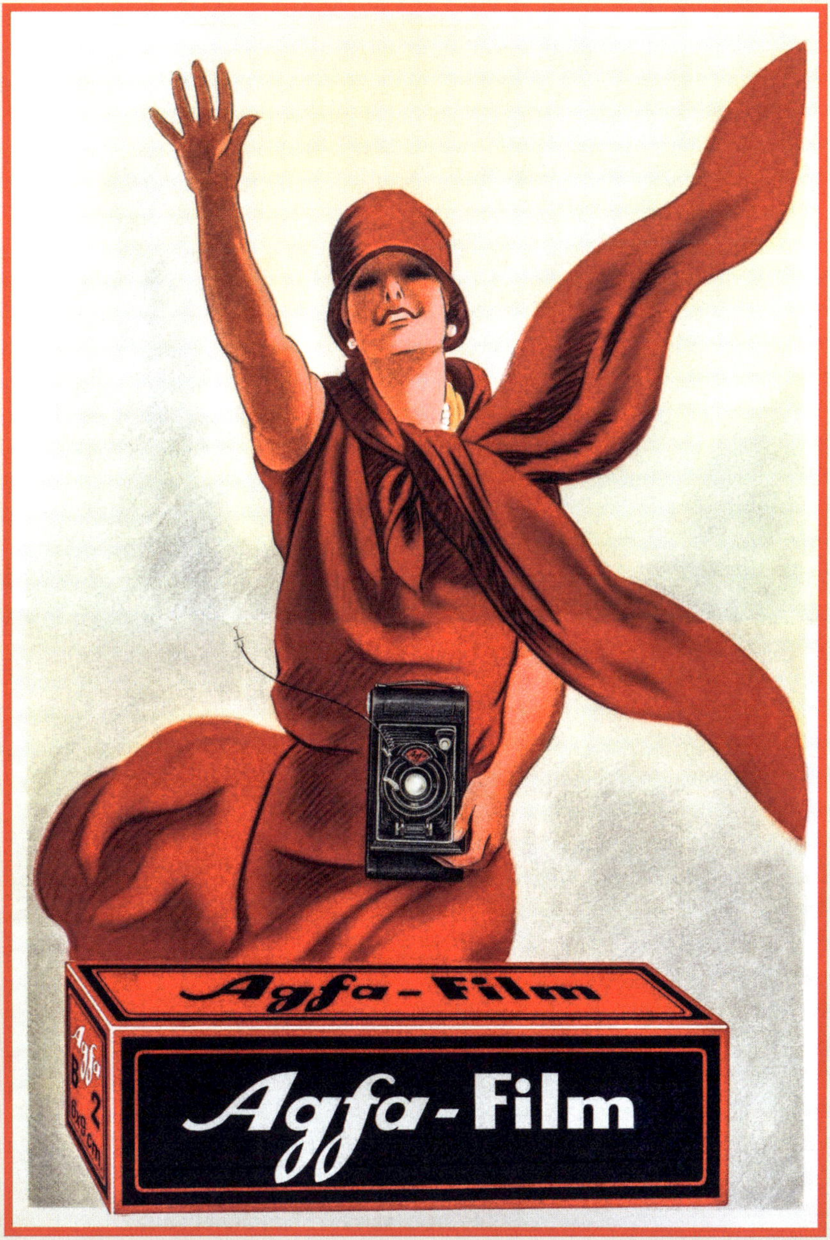
Agfa-Film
Agfa
B 2
6x9 cm
Agfa-Film

compete for market share. Zooming out, you can see some hotbed regions for photography popping up globally, with America, Germany, and Japan having some of the largest concentrations of historically important brands.

This chapter features fourteen notable brands that were responsible for driving innovation in camera and lens design, film formats, film formulas, photography chemicals, and other technologies. Many other companies have contributed to development and innovation in photography over the years (too many to include in a short book), but these fourteen are some of the bedrock brands, each leaving their mark in meaningful and unique ways.

Kodak

Perhaps no brand is more synonymous with film photography than Kodak. From the late 1800s to the early 2000s, Kodak changed how we communicate and document the world around us.

George Eastman had a dream: Make photography affordable and accessible to everyone with a medium so ubiquitous that it would "make the camera as convenient as the pencil." Having already patented a process for mass-producing dry plates, the real breakthrough came when Eastman perfected a cellulose solution that could be formed into a tough, transparent roll, which was then coated with a light-sensitive gelatin solution. The small, portable film rolls produced by this process could replace big, heavy plates that had been required up to that time. Camera designs followed suit.

The Eastman Kodak Company (formerly the Eastman Dry Plate and Film Company) was formed in 1892 following the release of the wildly successful Kodak No. 1 and Brownie cameras. The invention of roll film along with Kodak's business model ("You press the button, we

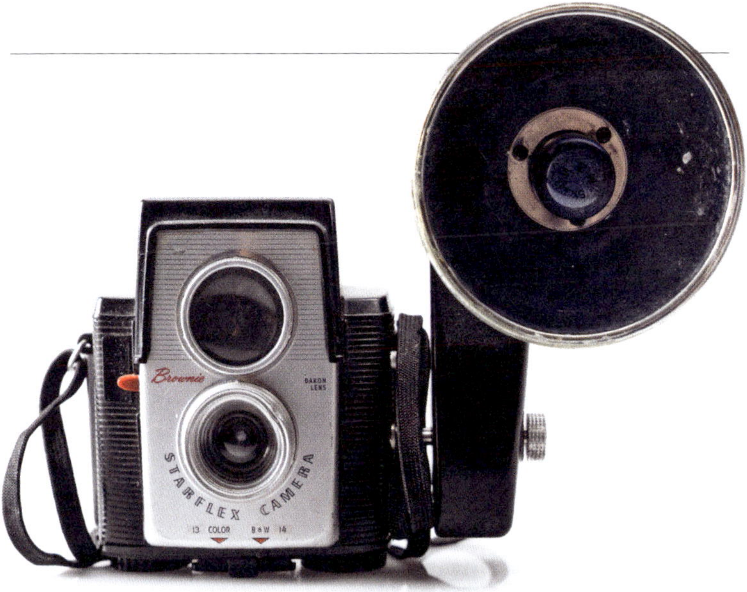

Kodak Starflex cameras were brought to market in 1957 as part of a number of "star" line cameras made to be low priced and accessible. Kodak sold 10 million cameras in that line during its production run.

do the rest") made photography easy and accessible to the masses seemingly overnight.

Kodak went on to drive photographic advancement through the Kodak Research Laboratories, which offered a range of innovations in film technology (for both consumers and motion pictures): chemicals for film development, cameras such as the Instamatic, machines for developing and printing images, film for medical imaging and some of the first X-rays, and even the world's first digital camera.

The company's impact was immeasurable, on society at large as much as on photography. Kodak's drive to put small cameras with short exposure times into the hands of as many people as possible changed

With their roots as an optics company, Voigtländer crafted and supplied various types of lenses for other camera manufacturers while also building a line of their own cameras.

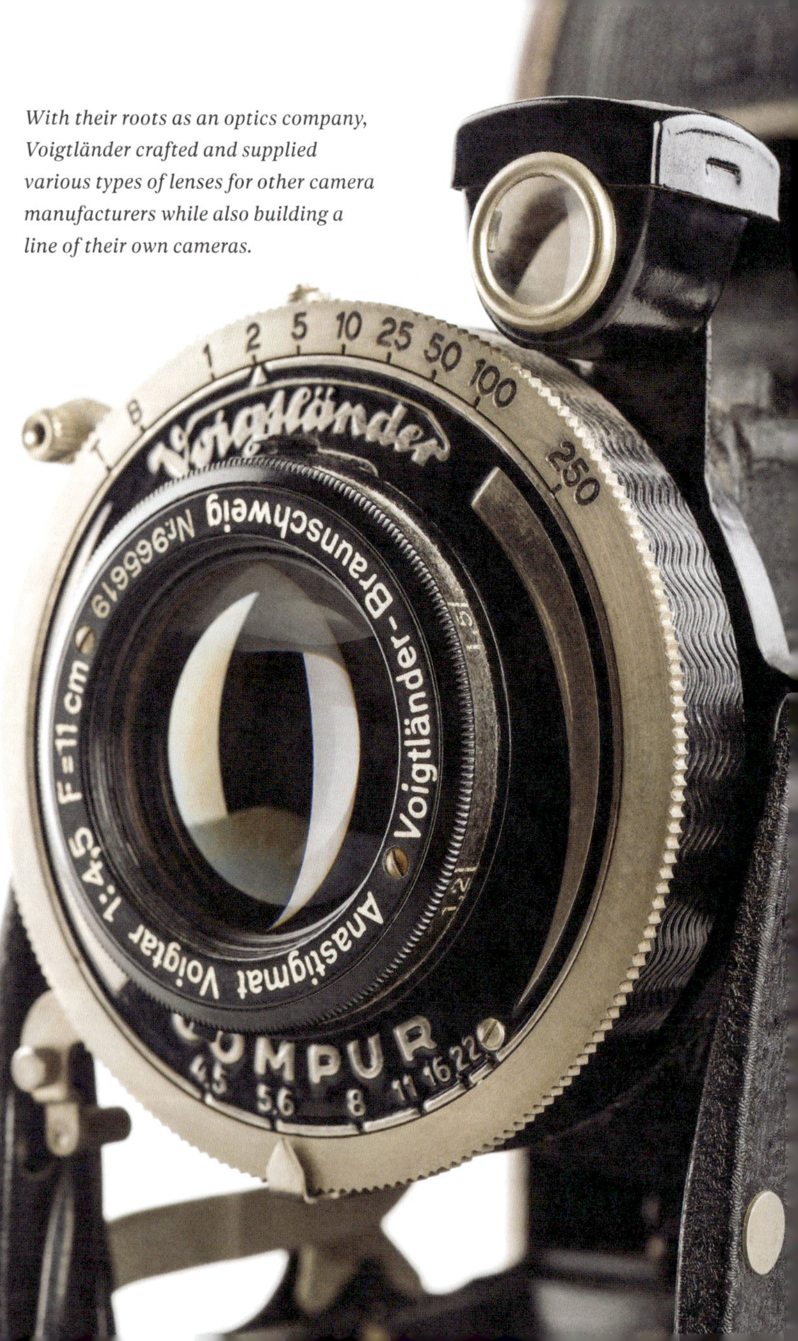

how we document and remember the world around us. Suddenly, the world went from stiff, formal studio portraits to informal photos of people smiling, laughing, and even jumping in front of the camera. Kodak film was on the first trips into space, documenting the moon landing and even the first Mars rover. Not only was Kodak an integral part of everyone's daily lives, but the company also helped document some of the most important events in modern history.

Voigtländer

Voigtländer is possibly the oldest camera brand, a company that helped shape many aspects of photography throughout its early development. Voigtländer was founded in Austria in 1756, long before the advent of photography as we know it. The company established itself as a pioneer, mostly focused on the research and production of precision optical instruments for navigation and surveying. They even produced opera glasses.

Their expertise in optics positioned them perfectly to contribute to photography early on. In 1840, just one year after the daguerreotype appeared, Voigtländer released a slew of photographic products that included an all-metal daguerreotype camera and, most importantly, the Petzval portrait lens.

Created by mathematician and inventor Josef Maximilian Petzval, this was the world's first objective lens made for cameras. It was faster than anything available at the time (f/3.5), which allowed for dramatically lower exposure times and was responsible for the entire portrait photography industry that it helped usher in.

Over the next few decades Voigtländer became one of the most influential photography brands in the world, producing high-quality lenses and cameras ranging from large-view cameras to rangefinders. Their innovative practices continued well into the twentieth century, and in 1959 the company released another historic invention: the Voigtländer Zoomar 36–82mm f/2.8, the first zoom lens for a 35mm camera. The lens was built around fourteen elements in five groups and had a minimum focusing distance of 4.5 feet (1.4 meter). It was big and heavy, but it was relatively fast, offering a consistent aperture of 2.8 over a useful zoom range, which eliminated the need for photographers to switch lenses to get the perfect shot.

The same year the Zoomar appeared, the Carl Zeiss Foundation bought controlling shares of Voigtländer, and six years later the company was integrated into Zeiss Ikon; the merger coincided with Voigtländer's release of the world's first 35mm compact camera with an integrated flash. The company has been bought and sold several more times as sales of film cameras have slid, but their historic impact on lens and camera design during photography's early evolution will never be forgotten.

Leica

Leica is firmly established in the pantheon of photographic brands with its high-quality optics, precise German engineering, and industry-changing rangefinder cameras. Founded in 1849, the company started as the Optical Institute in Wetzlar, Germany, a company dedicated to engineering high-quality optics and microscopes. In 1865 Ernst Leitz took over the company and changed its name to Ernst Leitz Optische Werke. Under his tenure, the company's microscope business attained global recognition due to his application of manufacturing skills he'd learned from the watch industry.

The company's foray into cameras started in 1914, all thanks to Oskar Barnack, an engineer who headed the company's development for cinema and measuring technology. Barnack was an avid amateur photographer with a personal stake in developing a lightweight alternative to the cameras of the time. Early in his career at the company, he introduced a prototype camera, the Ur-Leica (derived from the first three letters of "Leitz" and the first two letters of "camera"). The Ur-Leica was novel in its use of 35mm cinema film instead of the more common rolls of 120 film. And unlike cinema cameras, it shot horizontally, giving us the 24×35mm format that would become the standard for 35mm film cameras.

In 1924 Optische Werke, now under the leadership of Ernst Leitz II, made the bold decision to launch the Leica 1, the first commercially produced 35mm camera for consumers. A huge risk, but one that paid

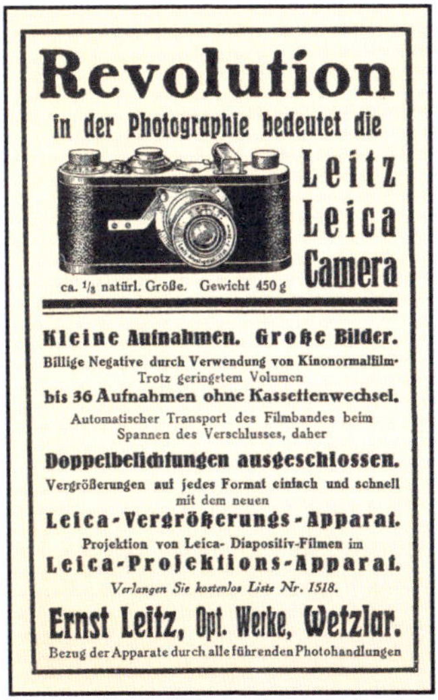

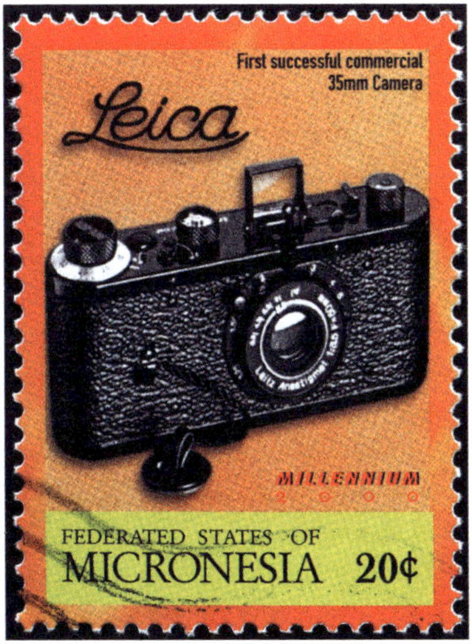

Left: In this advertisement for the Leica 1, the company announces a revolution in photography brought to you by the Leitz company, with thirty-six exposures on a 35mm film cassette. Right: Honoring the anniversary of Leica's release of the Leica 1, Micronesia released a stamp in March 2000.

off, the Leica 1 sold 56,000 units in the first five years of production, making the Leica 1 the chosen camera brand and format for some of the most important photographers of the time, such as Henri Cartier-Bresson and Robert Capa.

Leica went on to transform the photojournalism industry with their lineup of rangefinder cameras, particularly the M series, introduced in 1954 with the M3. Featuring a coupled rangefinder/viewfinder with parallax correction, a new lens mount, and quick film advance, the cameras gave photojournalists a small, lightweight camera that operated almost silently. Photographers could now work quickly and unobtrusively under fast-paced, hectic conditions. Iconic images such as Nick Ut's *Napalm Girl* from the Vietnam War and Robert Capa's *Death of a Loyalist Soldier* were taken with Leicas, and their rangefinders went on to set the standard of excellence that many brands tried to copy.

Hasselblad

From the wings of Swedish air force planes to the surface of the moon, Hasselblad's high-quality cameras have become iconic, finding their way into some of the most unique and extreme places in photographic history. Hasselblad is still considered one of the most desired brands for quality medium-format cameras.

F. W. Hasselblad and Co. was founded by Fritz Wiktor Hasselblad in 1841 as a trading company based in Gothenburg, Sweden. The company eventually created a photography division that was the exclusive importer and supplier of Eastman Kodak products for all of Sweden. This division expanded into a nationwide network of stores and labs, remaining through the generations under the ownership of the Hasselblad family.

During the height of World War II, the Swedish military shot down a German plane that contained an aerial camera. The government turned to Victor Hasselblad to design Sweden's own aerial surveillance cameras. Within months Victor had reverse-engineered the German camera and designed the HK7 under the newly formed company Ross AB, which eventually became today's Hasselblad.

Victor Hasselblad pivoted from war work to harness his experience and turn the company toward producing consumer-focused cameras. The first example released after the war was the Hasselblad 1600F, a 6×6 camera equipped with Kodak lenses, making use of the family's close relationship with George Eastman. The 1600F established the company's iconic square picture format and general camera design, creating the basis on which many future models were built.

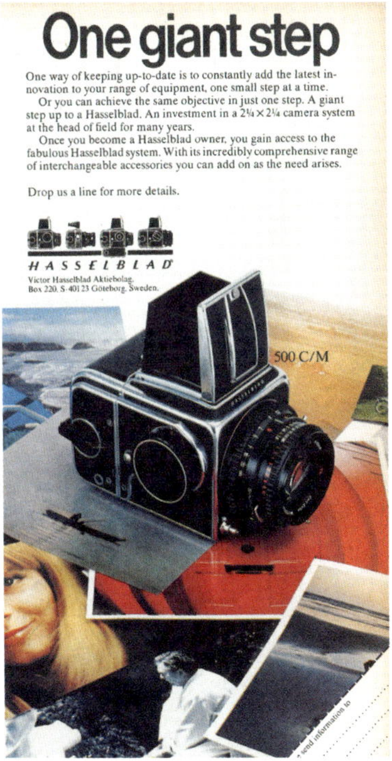

Leaning into their modular design, Hasselblad ads often included a row of silhouettes of various setups possible with their cameras.

Hasselblad cameras later found their way into space. Following the suggestion of astronaut Walter Schirra, an avid amateur photographer and Hasselblad user, NASA purchased and modified Hasselblad 500c cameras for use on the Mercury 8 space mission in 1962. To reduce their weight for space flight, the cameras were stripped of their auxiliary shutters, reflex mirrors, and viewfinders, and even had their leather exteriors removed. The modified cameras performed so well that NASA partnered with Hasselblad to produce cameras for the Gemini and Apollo missions that followed, including the moon landings. Famously, twelve Hasselblad cameras were left behind on the moon's surface.

The relationship with NASA helped skyrocket Hasselblad's popularity, and the company went on to set the standard for high-quality medium-format cameras throughout the rest of the twentieth century. That quality remains highly desirable among film collectors.

Zeiss

Founded in Jena, Germany, in 1846 by optician Carl Zeiss, this company began as a workshop for precision mechanics and optics, eventually producing microscopes that are considered some of the best after partnering with physicist Ernst Abbe. With the help of Abbe, the Zeiss Foundation produced lenses and optics for microscopes, still and film cameras, spectroscopy, eyeglasses, rifle scopes, medical devices, and more.

To compete with Kodak, in 1926 the Zeiss Foundation formed

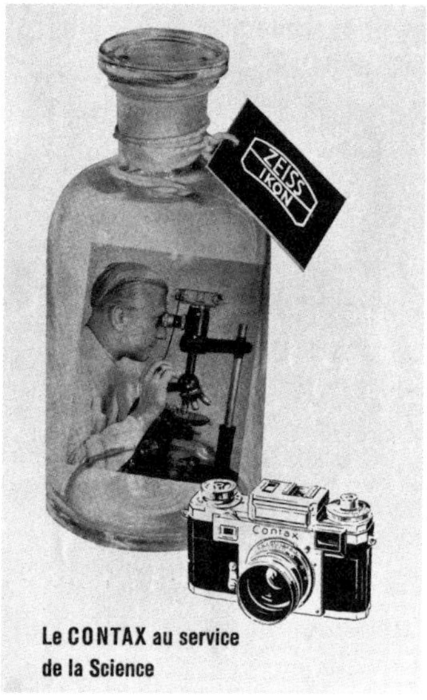

While Zeiss's roots were in optics, their Contax line of cameras brought to market reliable mechanical cameras paired with extremely sharp lenses.

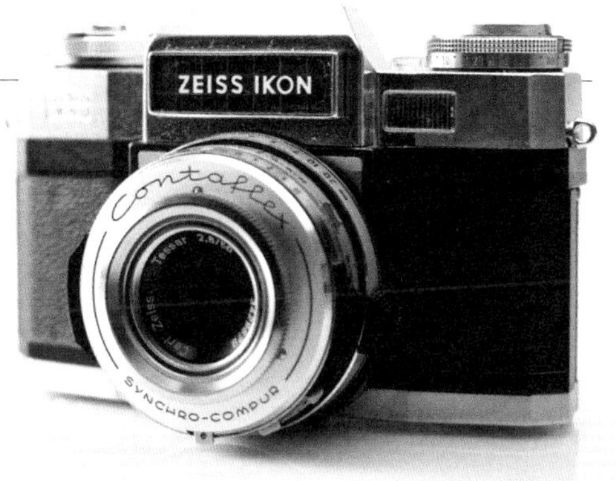

The Contaflex line from Zeiss Ikon were SLR cameras consisting of Zeiss optics fitted with the unique combination of a leaf shutter and an integral reflex mirror.

Zeiss Ikon by merging four major German camera manufacturers—Contessa-Nettel, Ernemann, Goerz, and Ica—and creating Europe's largest camera manufacturer. By combining the expertise and patents of these companies, Zeiss Ikon aimed to take on emerging companies from America and even other German companies like Leica.

In 1932 Zeiss Ikon released the Contax. Intended as a competitor for the up-and-coming Leica, the Contax was a rangefinder camera designed to use the same 35mm film used by Leica cameras, but they added a faster, vertically traveling metal shutter, which contrasted with the slower, horizontally traveling, rubberized silk shutters found in Leicas. Leveraging Zeiss's superior optics-manufacturing capabilities, the Contax was designed as a "system camera": a collection of lenses including wide angle and long focal length, something not available from Leica.

The aftermath of World War II had a profound impact on Zeiss, with Zeiss Ikon split into two manufacturing plants, one in East Germany and one in West Germany. This fragmentation eventually resulted in significant weakening of the brand. While the company continued to create high-quality cameras like the Zeiss Ikon Contaflex series in the 1950s, it faced growing competition from rising Asian brands like Canon, Nikon, and Pentax, who were able to make cameras nearly as well but at a much cheaper price point. Zeiss Ikon filed for bankruptcy and closed in 1972 and through an alliance with Yashica, their Contax line of cameras was spun off into it's own brand in 1975.

Zeiss's optical division remained strong, however, producing some of the sharpest, most sought-after camera lenses available for brands that included Hasselblad, Sony, Canon, Nikon, Yashica, and, of course, Contax. Their lens technology is found in high-end optical instruments and cameras today and continues to represent the highest level of precision German engineering.

Nikon

Nippon Kogaku K.K. was founded in Japan in 1917. The company that would later be known as Nikon followed a path similar to those of many of the German brands. As an optics manufacturer producing microscopes and other optical equipment, the company soon realized that if it was going to catch up to the quality of optics from the rest of the developed world, it would need some help. In 1921 Nippon Kogaku invited eight of the best optics engineers and scholars from Germany to Japan as advisors for their lens research and development teams. The guidance proved fruitful, and by 1931 the company was producing lenses that were optically competitive with Zeiss's Tessar lenses, thus paving the way for high-quality consumer camera lenses.

After World War II, the company added the production of consumer cameras to its other optical products, applying the brand name "Nikon" to its consumer camera line. Its first release, the Nikon 1, appeared in 1948. This camera featured a rangefinder with a design heavily inspired by Leica and Contax. The camera was relatively popular in Japan, but its nonstandard film ratio and incompatibility with Kodak films kept it from being imported to the United States. Nevertheless, the camera served as a launching point for Nikon's entrance into the consumer camera world.

Nikon's global reputation began to grow in 1950 when *LIFE* magazine photographers returned from covering the Korean War. The photographers

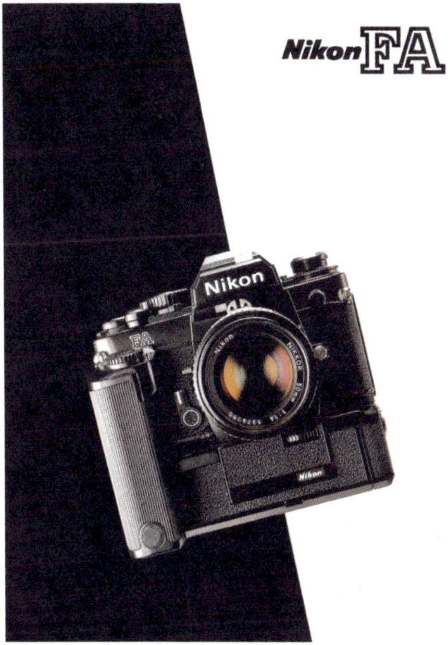

The Nikon FA, released in 1983, was the world's first camera with evaluative multizone matrix metering, leading the way for more complex metering systems in cameras.

praised the cameras for standing up to the bitterly cold weather they'd experienced; while other cameras were freezing up, the Nikons remained usable.

In the late 1950s Nikon abandoned development of their rangefinder cameras in favor of its F-series SLRs, which would go on to dominate the industry for the next thirty years. These models featured a system of interchangeable lenses, viewfinders, and motor drives. Nikkor lenses became known for their high image quality and robust builds, spurring many photojournalists to adopt the brand as their go-to cameras. They even found their way onto the space shuttle alongside Hasselblad. Nikon became a leader in film camera development and technology, introducing important advancements like electronic shutter control, and 1983's Nikon FA offered a camera featuring the world's first evaluative multizone matrix metering.

Nikon remains a foundational brand in global camera production, research, and advancement. It's one of the few traditional camera brands that survived the transition to digital photography.

Canon

Founded in 1933 in Tokyo, Seiki Kōgaku Kenkyūjo ("Precision Optical Instruments Laboratory") wasted no time developing and producing their first camera. Within a year they had a working prototype that was heavily influenced by the German 35mm rangefinders from Leica and Contax. The Hansa Canon included a viewfinder that popped up with the press of a button (playfully nicknamed the "surprise box"); the camera came mounted with a Nikkor lens from Nippon Kogaku, as Seiki Kogaku had yet to establish their own lens production factory. By 1949 Seiki Kogaku decided to change its name to Canon Camera Co., and the brand we know today was born.

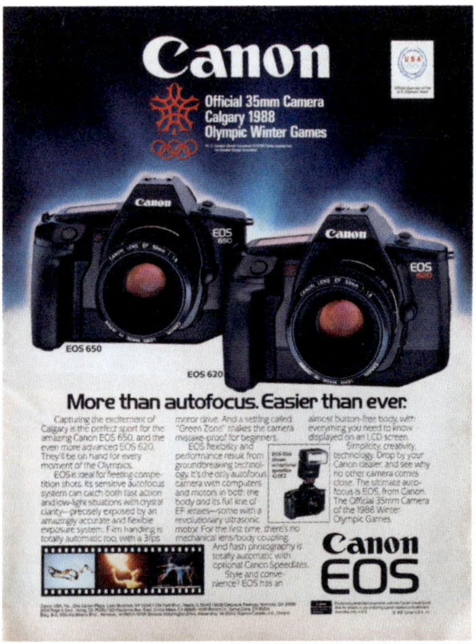

Canon debuted the EOS system in the Canon 650, designed from the ground up with the goal of offering a next-generation autofocus system.

From the 1950s through the 1970s, Canon released rangefinder and SLR cameras with innovative features. These included the Canon Pellix in 1965, which included a novel pellicle reflex mirror that was semitransparent and stationary. By the 1960s Canon had shown that Japanese manufacturers could produce competitive cameras that could pace German cameras not only in quality but also in innovation.

Their capacity to integrate electronics into their products solidified Canon as a powerhouse brand. In 1976 Canon launched the well-known AE-1, a camera designed with a microprocessor that controlled everything from the electromagnetic shutter to the metering, aperture, and exposure. The AE-1 was hugely successful, selling over 5.7 million units and demonstrating the possibilities of microprocessor integration.

In 1987 Canon launched its Electro-Optical System (EOS) with the debut of the Canon EOS 650. This was a watershed moment for the company, and arguably for film photography across the board. With their EOS system, Canon deeply integrated electronics into the heart of the camera, introducing a new Electro-Focus (EF) lens mount that removed all mechanical linkages and relied solely on electronics to allow the camera and the lens to communicate with one another. The system offered many features, including motors within the lens to drive autofocus much more quickly than motors within the camera body; through-the-lens (TTL) metering integrated with external flashes; and even eye-controlled autofocus, introduced in 1992 with the EOS 5.

The EOS system made Canon one of the largest camera brands in history. Most other major manufacturers followed suit, redesigning their camera systems to complete. EOS has remained at the heart of Canon cameras down to the present.

Mamiya

While Mamiya produced cameras across a variety of formats, the company's reputation was forged within the medium-format industry with their TLR and SLR cameras. As with many other Japanese manufacturers, Mamiya offered great cameras at more affordable price points compared to many Western competitors.

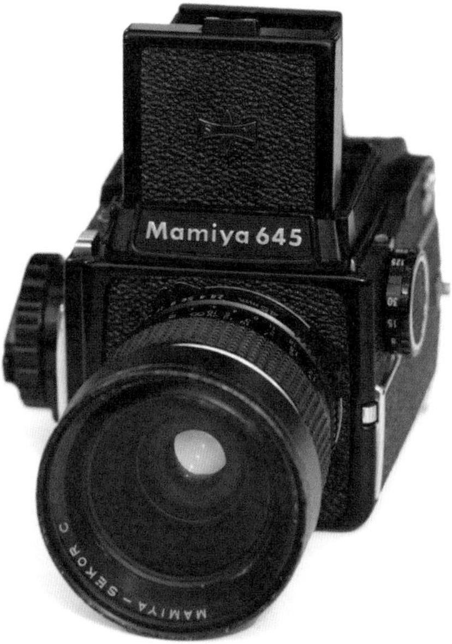

The Mamiya 645 line of cameras was a series of modular medium-format cameras released in 1975, offering a quality, lower-cost alternative to other modular systems, such as what Hasselblad offered.

Mamiya Optical Works was founded in Tokyo in 1940 by camera designer Mamiya Seiichi and Sugawara Tsunejirō. At first the company offered only one camera: the Mamiya 6, a 6×6 medium-format rangefinder with a coupled viewfinder and a folding, retractable lens. By 1948 the company's offerings were more varied, with the Mamiyaflex lineup of 6×6 TLRs, a 35mm rangefinder (Mamiya 35), and a set of subminiature 110 cameras in the Mamiya 16.

In 1957 Mamiya introduced two cameras with a design feature that would define the brand moving forward: modularity. The Mamiya Magazine was a 35mm rangefinder with removable, interchangeable film backs, while the Mamiyaflex C was a 6×6 TLR with interchangeable lenses that introduced a more affordable medium-format competitor to brands like Hasselblad.

With the introductions of the RB67 and the Mamiya M645 in the 1970s, Mamiya solidified its position as a medium-format manufacturer for professional and serious amateur photographers. The RB67 and

M645 marked the beginning of a new series of highly modular 6×7 and 6×4.5 camera systems, with accessories ranging from different film backs to automatic film winders, lenses, viewfinders, and more. A bigger camera initially designed for studio photography, the RB67 featured a rotating film back that allowed the user to change film orientation without moving the camera. The M645 was smaller, lighter, and designed to be handheld. And because the resulting images were slightly smaller, it was more economical to shoot with, fitting sixteen images onto a roll of 120 versus the RB67's ten images per roll.

The late 1980s and early 1990s saw the introduction of modernized Mamiya 6 and Mamiya 7 rangefinders that shot 6×6 and 6×7, respectively, offering beautiful, large negatives in a compact body. The RB67, RZ67, and 645 series, along with the Mamiya 6 and 7, still hold their own in terms of image quality and remain highly collectible.

Asahi Pentax

The Asahi Optical Joint Stock Co. opened in November 1919 in Otsuka, a neighborhood of Tokyo. The company began as a lens manufacturer for eyeglasses and binoculars but, like many companies that appeared in this time, it saw its manufacturing specialties change during World War II with the shift to producing optical equipment for the government. After the war, Asahi took a calculated risk that paid off, applying experience manufacturing cine and camera lenses to produce its own camera. In 1952 the company announced the release of the Asahiflex, Japan's first SLR.

In 1957 the company released the pivotal Asahi Pentax, the first SLR to feature a film-advance arm and a shutter-speed knob on the right side of the camera, a pentaprism viewfinder in the center, a folding film-rewind crank on the lefthand side, and an automatic return mirror. If that description sounds familiar, it's because this design layout

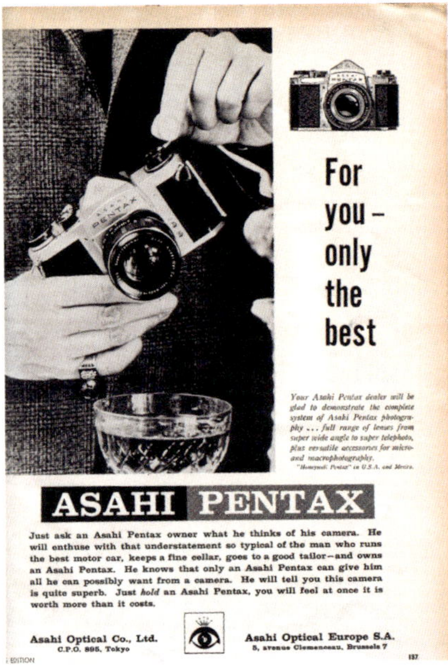

While sometimes overshadowed by Nikon and Canon, Pentax released models that rivaled the quality of the larger Japanese brands.

became standard for almost all other SLRs in the following decades.

Pentax's innovation continued in 1960 when the company debuted a working prototype of the world's first SLR capable of taking TTL light meter readings. The camera was released four years later in 1964 as the Pentax Spotmatic, deriving its name from the automatic spot meter in the prototype (the production camera swapped the spot meter for an average meter for more consistent results). The Spotmatic lineup continued, adding innovative features through the years, including the release of the Pentax Electro Spotmatic in 1971, which was the first SLR to feature automatic exposure with an electronic shutter.

Throughout its reign, Pentax was the leader of SLR production. Nine years after releasing its first camera, in 1966 Asahi became the first camera company to produce its one-millionth SLR, and it would take the company a mere three more years to double that mark. Not only did Pentax drive innovation, their cameras, such as the legendary K1000, were known for their durability and affordability, helping to democratize access to SLR film cameras for generations.

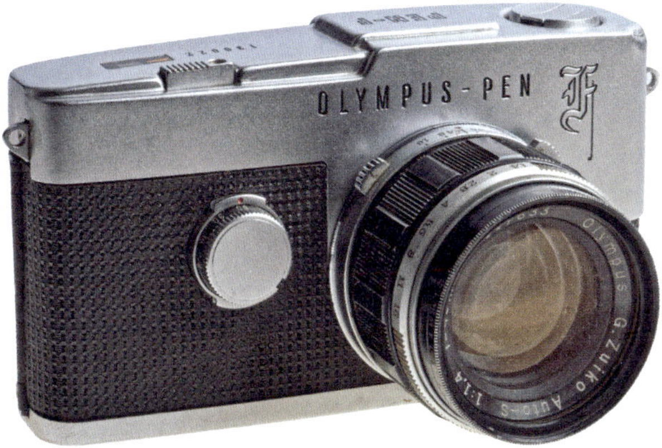

The Pen F was Olympus's successful half-frame, interchangeable SLR, allowing for seventy-two images to be taken on a thirty-six-exposure roll of film.

Olympus

Olympus's photographic legacy is built on the company's constant push for technological and design innovation and calculated risk-taking. From the introduction of the smallest (at the time) 35mm SLR to the invention of off-the-film metering, incredibly high-quality lenses, and even new film-size formats, the company constantly pushed the boundaries of camera technology.

Its origin story is similar to those of many other camera companies. Founded in Tokyo in 1919 as Takachiho Works Co., Ltd., the company began as a microscope and optics manufacturer, eventually becoming the official microscope supplier for the Japanese navy. In 1935, attempting to diversify its products, the company started a lens division named Zuiko, initially making lenses based on the German Tessar design.

Having a difficult time finding enough clients to buy lenses, Olympus decided to produce its own cameras featuring its Zuiko lenses and, in 1937, released the Olympus Standard and the Semi Olympus II. Olympus would go on to make a variety of cameras, from TLRs to rangefinders, SLRs to subminiature cameras, but two lines had a sizable and lasting impact on the photo world: the OM series and the Pen series.

The Pen camera was released in 1959. Designed by Maitani Yoshihisa, this half-frame camera shot on regular 35mm film, providing double the number of exposures on a roll of film. It was small and light and could be taken anywhere, like a pen (from which its name was derived). The first Pen camera was simple, offering no light meter, only five selectable shutter speeds, and a 28mm prime lens, but it would be the first in a long line of successful cameras for Olympus, selling over 17 million units before production ceased in the early 1980s.

The OM series launched in 1972 with the OM-1. Smaller and lighter than any SLR on the market, the series introduced unique, innovative features throughout its lifetime, such as off-the-film light metering in the OM-2, which measured light reflecting off the surface of the film, a feature that allowed for metering to continue even while shooting. The OM (film camera) series remained a bedrock of Olympus cameras for decades, going through a number of models and design revisions over the years and finally ceasing production in the early 2000s.

Rollei

Like many of their German camera manufacturing counterparts, Rollei is known for high-quality, precision manufacturing and for setting a standard for supremely well-built twin-lens reflex cameras. When you think of TLRs, you most likely think Rollei. And if you don't think of Rollei, there's a good chance that the brand you do think of copied them.

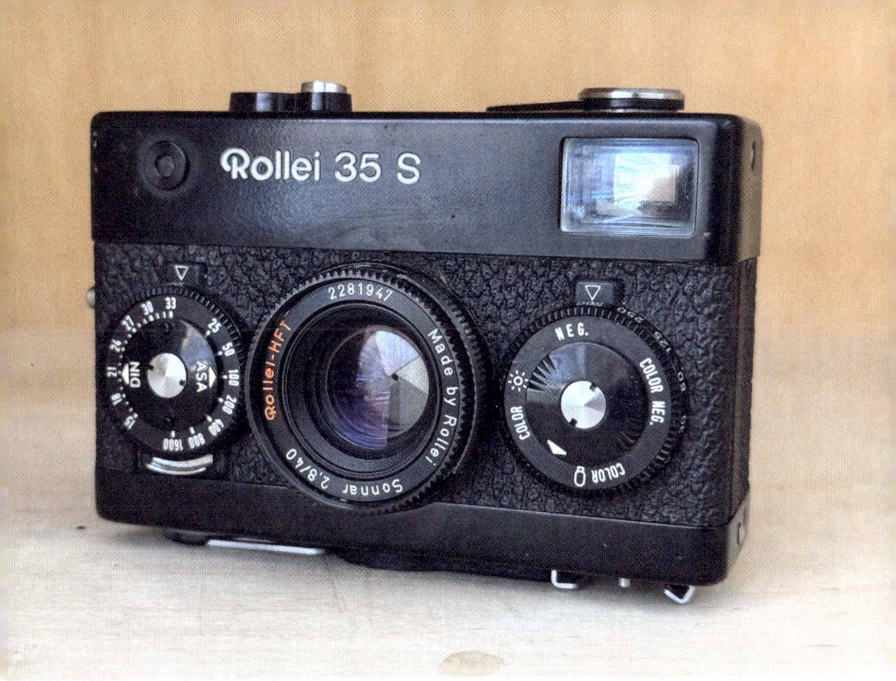

While best known for their TLR line of Rolleiflex cameras, Rollei also released a highly portable 35mm camera, the Rollei 35, which has developed a cult following over the years.

The company was founded in 1920 by ex-Voigtländer employees Reinhold Heidecke and Paul Franke as Franke & Heidecke, and their mission of creating a high-quality TLR finally came to fruition in 1927 with the release of their first Rolleiflex prototype. To ensure the camera could withstand the rigors faced by photojournalists operating in harsh environments, the camera was designed with no leather bellows or cloth shutters—everything was metal. To achieve this, the Rolleiflex deployed some novel design features: a lens board that slid against the frame of the camera to form a light-tight seal; a rack-and-pinion system that moved the lens board, allowing for precise focusing without the need for biodegradable bellows; a crank-lever film advance with an exposure counter; and high-quality German optics from either Zeiss or Schnieder. After a few more design iterations, in 1929 the camera went into production and was released to the public. It was so successful that, for the first few years, demand outpaced production capabilities, resulting in an expansion during the Great Depression, something not many companies could claim. By the 1950s the Rolleiflex was considered the press camera of choice for photojournalists and had arguably set the standard for TLR design.

As pressure mounted from Japanese brands and the market began to shift toward the more portable 35mm SLRs, Rollei broadened their offerings. Starting in the 1960s, this included medium-format SLRs, 35mm SLRs, and subminiature cameras, all with high-quality optics and many with industry-leading technology. But these offerings never took off like the company's TLRs, and sales continued to decline until eventually the company filed for bankruptcy in 1981.

Today Rollei produces high-quality film, paper, and darkroom chemicals, but their true legacy remains rooted in the mark they left on the TLR landscape.

Fujifilm

What started as a small subsidiary of a chemical company in Japan grew to become a true competitor for Kodak and Polaroid in the film space. At present Fujifilm is a leader in film production and digital photography. Developing a wide selection of color film known for its distinct color quality and innovative offerings like Fuji Velvia and Fuji Instax (an instant film to rival offerings from Polaroid), Fuji grew into a cornerstone company in the photo industry, producing products that rivaled and bested the great photographic film brands—and outlasted them all.

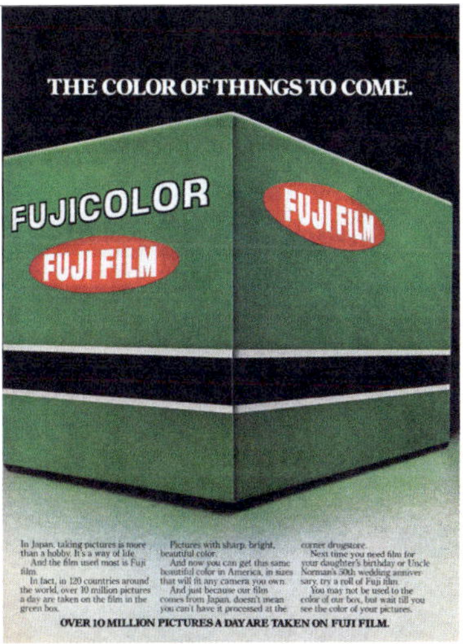

While the Fujicolor process was introduced in 1955, this 1978 ad in People *magazine has Fujifilm proudly proclaiming that over 10 million photos a day are shot on Fujicolor film.*

Founded in 1934 as a subsidiary of Dainippon Celluloid, Fuji's foundational mission was to produce still photography and motion picture film domestically, eventually expanding to include dry plates for graphic arts, photo papers, and X-ray film. In 1940, in an effort to expand its offerings beyond photosensitive materials, Fujifilm opened an optical factory, manufacturing optical glass and camera (Fujinon) lenses.

In 1948 Fujifilm released both its first color film, named Fuji Color Film, and its first camera, the Fujica Six, a 6×6 medium-format folding camera whose name derived from the combination of Fuji and "ica" for "camera," à la Leica. It was a turning point for the company that would spur the development of a slew of cameras and film, as seen just a few years later when it released Neopan SS, a black-and-white ortho-panchromatic film that would rival Kodak Tri-X and become a Fujifilm classic.

Not only was Fujifilm a driver in moving technology and products to market, but those products were responsible for helping change the community and culture around photography. One of Fuji's more underrated contributions to photography came in 1986 with the release of the Fuji QuickSnap, the world's first one-time-use, disposable camera. For the first time you could purchase a camera that cost slightly more than a roll of film and not worry about losing or breaking it. Cameras could be handed to children on vacation or guests at parties to document the experience. The idea of sharing and community also included Fuji's line of instant film released in the 1980s and then rebranded in 1998 as the Fujifilm Instax line of cameras and instant film still made today.

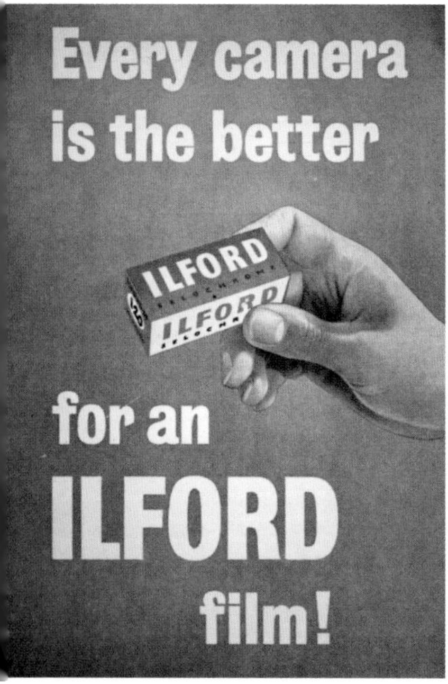

Emerging as a legitimate competitor to Kodak film stocks, Ilford offered a number of analog products that included film, paper, and chemicals.

Ilford

Like Kodak, Ilford is synonymous with black-and-white film photography. For over a century, Ilford was at the forefront of photographic film and paper R&D, creating hallmark products like its HP5 black-and-white film and multigrade darkroom papers. Ilford continues to produce photographic products nearly 150 years after it was founded in 1879 by Alfred Harman as Britannia Works, making dry plates for photographers. In the early 1900s the name was changed to Ilford after the neighborhood in East London, and business expanded to manufacturing bromide and alpha paper as well as roll film starting in 1912.

Ilford's business expanded throughout the early twentieth century, pushing research and production in new roll film, dry plate, photo paper, and camera technologies. While many in North America came to know Ilford through their film and photo paper offerings, Ilford cameras at its height accounted for over half of the UK's 35mm camera market thanks in large part to their affordable Sportsman line. Manufactured by German camera maker Dacora, the Sportsman offered a cheaper alternative to the popular Kodak Retina line in the 1950s and 1960s.

In the darkroom Ilford continued to innovate with the introduction of multigrade (variable contrast) paper in the 1940s. This first-to-market paper allowed photographers to use up to four filters to adjust contrast while printing in the darkroom. Slow to take off, by the 1970s their latest iteration offered far better contrast control, capable of using eleven different filters when printing and becoming a beloved black-and-white paper among film photographers.

In 1931 Ilford introduced hypersensitive panchromatic plates, later shortened to HP. Four years later that formula was brought over to roll film. The original HP had an ISO of 160, and the name went on to form the basis for a line of film: HP, HP2, HP3, HP4, and HP5, with HP5 Plus being the last film in the family to be released in 1989.

Today Ilford operates under the ownership of Harman Technologies, which also owns and operates the film and paper company Kentmere. The company recently launched its own film lines, Harman Phoenix and Red.

Agfa

Formed in 1867 as Aktien-Gesellschaft für Anilin-Fabrikation in Rummelsburg, Germany, Agfa (for short) initially was founded as an aniline dye company. In 1887 it entered the world of photography after its chemists suggested exploring photographic chemical production.

In 1891 the company introduced its famous Rodinal developer, which came as a liquid concentrate rather than the typical powder chemicals offered by Kodak and other companies. Rodinal became a favorite among many black-and-white photographers and a staple in darkrooms for over a century. Rodinal is still available today, making it one of the oldest continuously produced photo products in history.

In an effort to expand its production capabilities, in 1910 Agfa opened a film-production factory in Wolfen, Germany, a move that made the company a major player in the photo industry. At the time it was one of the world's largest film factories, second only to Kodak's factory in Rochester, New York. Agfa continued to push innovation with their research into color film in the 1920s, and in 1932 Agfacolor was introduced as a competitor to offerings from Technicolor and Kodak. What made Agfacolor so unique was the way its dye-couplers were integrated into the film, unlike Kodak's Kodachrome; this innovation made Agfacolor cheaper and easier to produce and for labs to develop.

Agfa produced several cameras throughout its history, including rangefinders, box cameras, and even subminiature cameras using 110 film. Of all the models Agfa produced, one in particular left a historic mark: the Optima.

Released in 1959, the Optima was the first camera manufactured with automatic programmed exposure. A large selenium light meter in the front of the camera produced a small amount of power in response to how much light hit it. This electrical current was used to set the aperture and shutter speed, removing the need for the photographer to make any manual adjustments. The Optima became an entire line that was produced into the 1980s.

While its camera production eventually ceased, Agfa's film and chemical production continues today. You can find Agfa films produced and branded for companies like the Lomography Society International and Rollei.

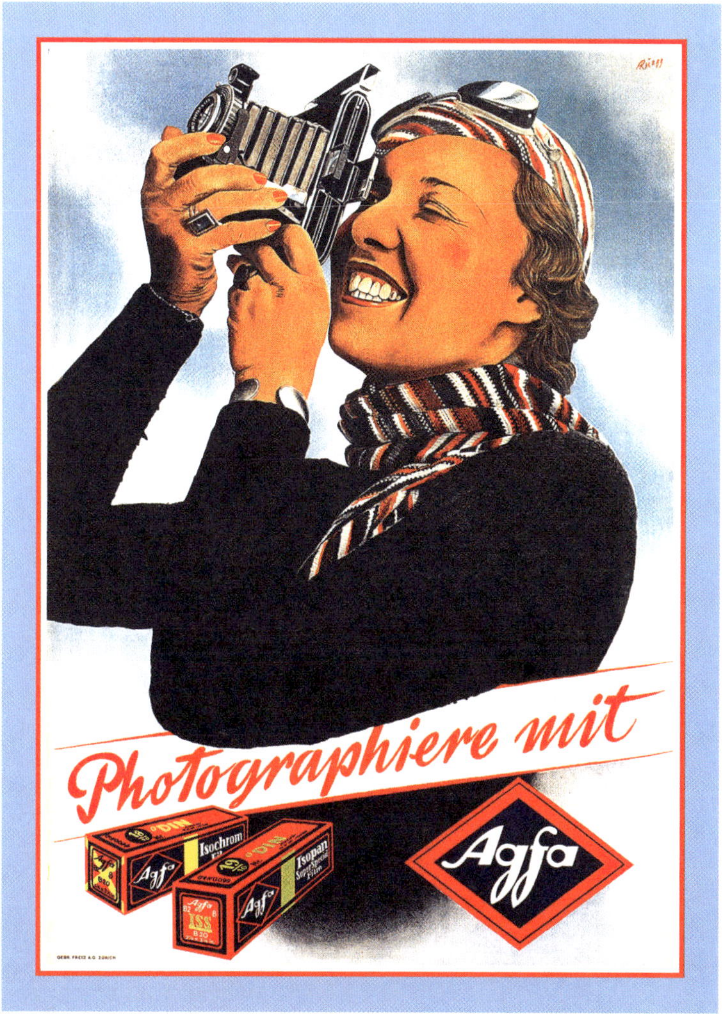

Getting their start in the dye manufacturing world, Agfa pivoted into photographic materials to offer a number of film stocks and darkroom chemicals for photographers.

CHAPTER FIVE

The Legends

Historic Film Cameras

THROUGHOUT THE HISTORY OF FILM CAMERAS, certain models stand out for introducing something new or changing the direction of the market. In this chapter, we explore nineteen cameras that have had an impact on photography, some even on society. From models that democratized photography for the masses to those that helped document wars, famine, and city streets, to cameras that recorded our first steps on the moon, cameras have shaped how we see the world and how we communicate with one another. Once expensive luxuries, cameras are now deeply integrated necessities of our daily lives. These are the film camera legends.

Practice makes perfect: Neil Armstrong practices using a Hasselblad camera (on Earth) while in a pressurized spacesuit.

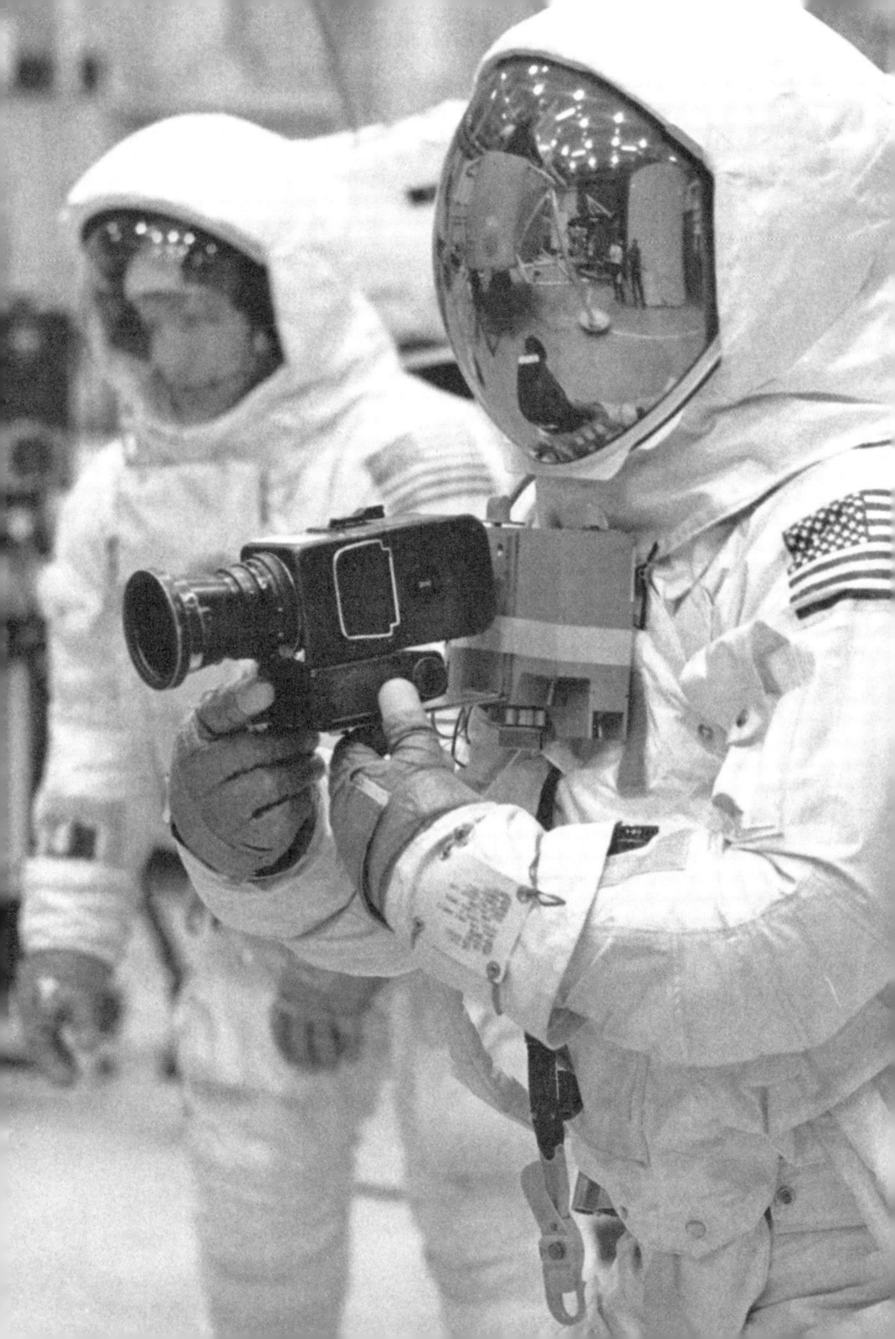

Kodak No. 1

Boxy, simple, utilitarian, yet accessible and easy to use—the Kodak No. 1 was truly a watershed moment not only for Kodak but for photography as a medium and an industry. Gone were the days when you needed to invest in a home darkroom, buy and store expensive photographic plates and chemicals, and learn the nuances of operating the camera and of the chemistry of film development.

The Kodak No. 1 resulted from a collaboration between George Eastman and Frank A. Brownell, a Canadian cabinetmaker, so its construction with a wooden body should come as no surprise.

Operationally, the camera was quite simple. The Kodak No. 1 had a fixed aperture (f/9), a single shutter speed (1/25), and a cord to cock the shutter. Once an image was taken, a key atop the camera advanced the film. There was no viewfinder or ground glass of any sort, just a V indented into the leather on top of the camera that gave the photographer an approximate field of view.

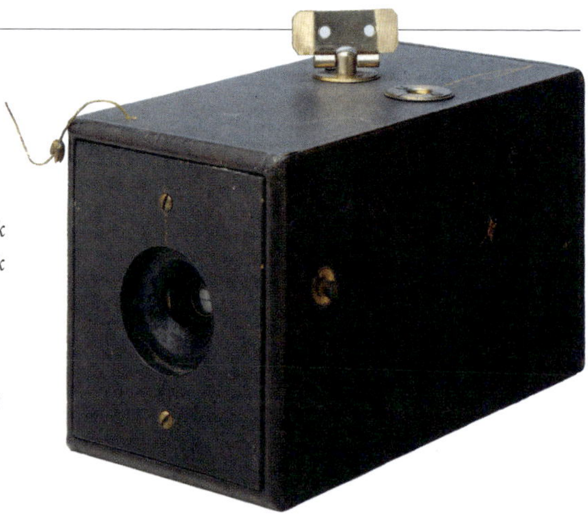

Designed by cabinetmaker Frank Brownell, the Kodak No. 1 had more visual elements in common with functional furniture than later cameras would have.

The camera produced 2⅝-inch (6.7-centimeter) circular images on Eastman stripping negative film that was preloaded in the camera. One hundred images could be shot on the roll before the camera was sent back to Kodak, where the roll was developed and printed, and the camera reloaded and returned—all for $10. The No. 1 was truly one of the world's first point-and-shoot cameras.

The release of the Kodak No. 1 in 1888 was the start of the democratization of photography. But at $25 for the camera, the No. 1 was still well out of reach of the typical working-class American. In many ways the Kodak No. 1 can be seen as a proof of concept—a test run for a camera and a process that would be implemented later at a more accessible price point. And that's exactly what happened years later, with the release of the even cheaper (that is, more accessible) Kodak Brownie.

Kodak Brownie

Released in 1900, the Kodak Brownie brought photography to the masses. Building on the mission set by the Kodak No. 1, the Brownie was also designed by Frank A. Brownell, and the two cameras share many of the same design language.

The Brownie shot on Kodak 117 film and introduced the world to the 2¼-inch (5.7-centimeter) square image. A simple box camera built out of cardboard and covered in leatherette with a detachable film-winding key and no built-in viewfinder, the camera debuted with the cost of $1, making it far more attainable than its predecessor.

Originally marketed for children, the name Brownie was a nod to the Brownie cartoons featuring fairy-like creatures created by Canadian artist Palmer Cox. Early advertisements actually featured these cartoons on the camera. But it wasn't just children buying and using the Kodak Brownie: In just five years, 10 million units were sold, con-

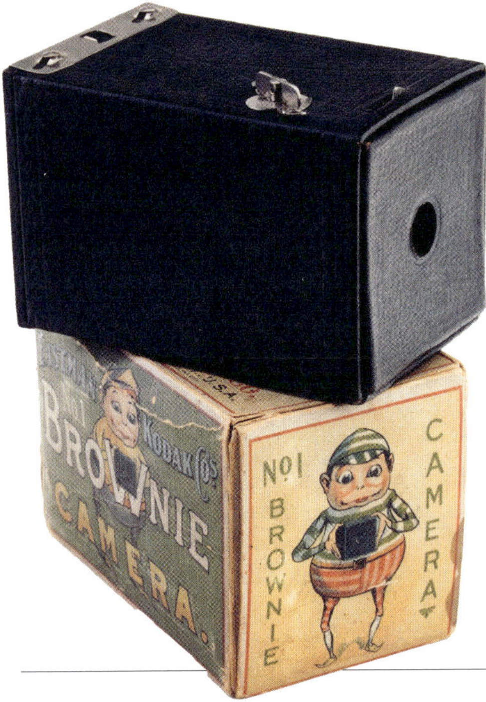

While most lenses contain several glass elements, the Brownie's lens contained just two pieces of glass and a rotating shutter, allowing Kodak to keep the camera affordable.

vincing Kodak that the camera should be marketed as an entry-level camera for the masses.

The Brownie evolved into an entire line of cameras that included box cameras, folding cameras, and rangefinders, ending in 1986 with the subminiature Brownie II that used 110 cartridge film.

Responsible for taking the camera from luxury item to everyday accessory in the hands of millions of people, the Brownie is arguably the most important film camera in photographic history.

Leica 1

The Leica 1, the first production camera to use 35mm film, was introduced to the world in 1925 at the Leipzig Spring Fair. Little did onlookers know that over the next century the vast majority of consumer film cameras sold would be 35mm cameras. In fact, 35mm became such a standard that when the transition to digital cameras occurred, film sensors considered "full frame" kept the same 24×36mm aspect ratio established by the Leica 1 in 1925.

The Leica 1 was the brainchild of Oskar Barnack, a researcher and head of development for cinema and measuring technology at Ernst Leitz Optische Werke in Wetzlar, Germany. Barnack was an amateur photographer looking for a lightweight portable camera that he could carry without aggravating his asthma. Combining rolls of 35mm cine film and lenses he obtained at Leitz, he came up with the prototype that became the Leica 1.

The Leica 1 was a rangefinder camera equipped with a focal-plane shutter capable of speeds of 1/20 to 1/500 of a second.

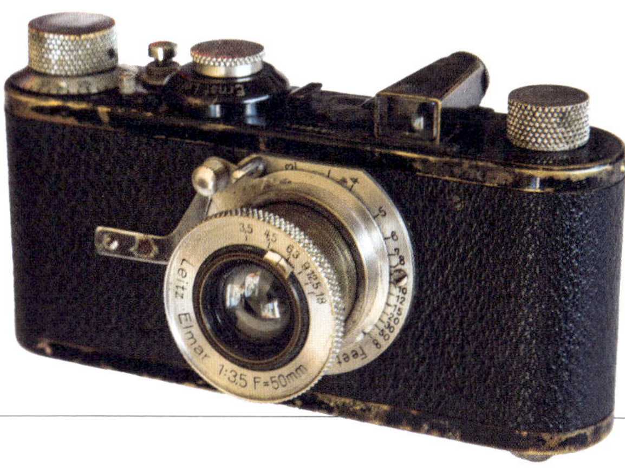

The Leica 1 shares the same film-winding knob as later Leica cameras, making the film rewinding process a guaranteed workout for a photographer's wrist.

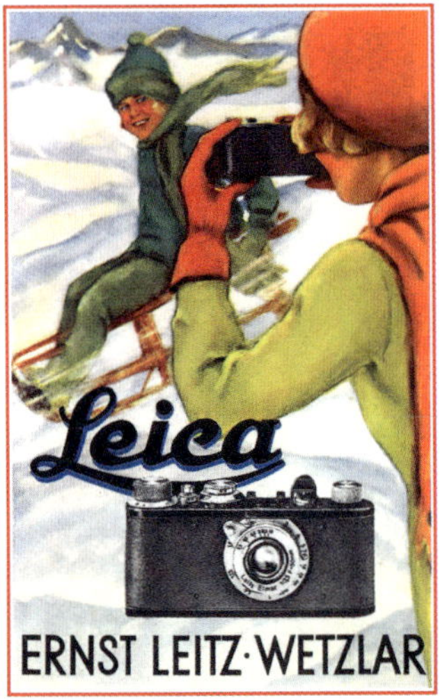

Many of Leica's advertisements from this period consisted of hand-painted scenes of people documenting daily life, showing how portable the camera was when in use.

The early lens prototype was based on the Cooke triplet of 1893, but shooting on a small negative and then enlarging to a desired print size necessitated a lens capable of high resolution in order to create quality images. In 1925 a breakthrough in lens technology in Leitz's eyeglass division allowed for the implementation of higher-quality optics, and a custom fixed 50mm f/3.5 Elmar lens with an aperture range of f/3.5–f/18 was designed just for the Leica 1.

The Leica I was replaced by the Leica II in 1932, but not before over 57,000 cameras were produced with various updates made to the lens offerings. The original design served as the base from which all future Leicas evolved.

Kine Exakta

Manufactured by Ihagee Kamerawerk Steenbergen GmbH in Dresden, Germany, the Kine Exakta holds the title as the first 35mm production SLR. Released in 1936, it featured design elements you would expect from a 35mm SLR—and many you wouldn't. Machined out of aluminum, it was built like a tank and had a price tag that kept it out of range of all but professional photographers and serious amateurs.

One immediately apparent design feature is the camera's waist-level viewfinder. Designing a useful waist-level viewfinder for a film format this small required a novel design. Because of the small size, it was nearly impossible to reliably tell if your image was in focus using the typical ground glass found in other waist-level finders, so engineers opted for a plano-convex magnifying glass, with the convex side facing the photographer and the flat side ground into a focus screen. An additional magnifying glass swung out for fine-focus work, further helping the user focus at waist level. To help project a bright image in the finder and achieve fast focusing, the camera came equipped with a fast Zeiss Tessar f/2.8 lens and later an even faster f/2 Schneider lens. The shutter was a focal-plane type made of cloth that traveled horizontally and offered speeds capable of 1/25–1/1000 of a second, along with a bulb mode.

Many of the camera's functions were reversed when compared with what would be found on most later SLRs. The film was loaded into the right side and traveled to a pickup spool on the left. The

While the Kine Exakta comes with a waist-level finder, composing shots can also be accomplished by using the sport finder found in the lid of the finder cover when opened.

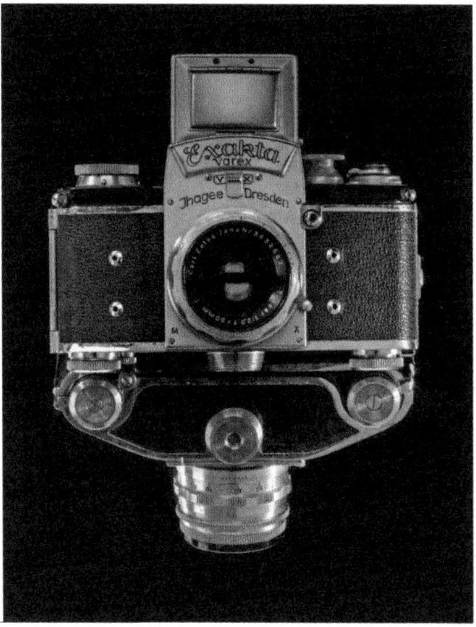

film advance was a knob on the top lefthand side of the camera that could be operated by the user's left thumb. Rotating the knob not only advanced the film to the next shot but also cocked the shutter and advanced the shot counter. Shutter speed was also set with the knob by lifting it up and rotating to the desired speed.

One of the camera's more unique features was an optional film-cutting blade. The design of the camera allowed for the film pickup spool to be installed into a cassette; as images were shot, the exposed film was unloaded from one cassette and loaded into a second cassette on the opposite side. The blade allowed the photographer to stop shooting mid-roll, cut the film off, and remove the cassette of exposed images to be developed without shooting an entire roll of film.

The Exakta line continued production with various models through the early 1970s.

Graflex Speed Graphic

There's a classic "scene of the crime" moment in film noir movies of the 1940s and 1950s. It's on a street, it's raining, and the press and police are gathered around a body. Everyone's wearing long trenchcoats and fedoras. The journalists wield 4×5 "press cameras" with a big flash attached to the side. That large camera was most likely a Graphic model by Graflex, which for decades was the de facto press camera for most journalists.

What's known as the "Speed Graphic" is actually a line of models that include the Speed Graphic, Pacemaker Speed Graphic, Pacemaker Crown Graphic, Century Graphic, Super Graphic, and Super Speed Graphic. Produced from 1912 to the early 1970s by the Graflex company in Rochester, New York, these cameras were designed to be workhorses for journalists, though professionals from many other fields also used them.

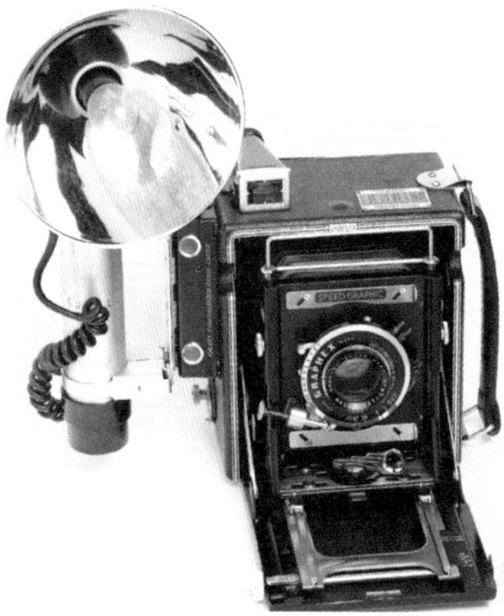

While certainly not small by today's standards, the compact form paired with its large negatives made the Speed Graphic a favorite among journalists.

Graphic cameras are large-format field cameras with several unique features. The Speed Graphic came with a focal-plane shutter capable of speeds as high as 1/1000 of a second (hence the "speed" in its name) and came mounted with a leaf-shutter lens. The camera had *three* viewfinders, which meant it could be used as a view camera for composing an image projected onto the ground glass in the rear of the camera, or as a rangefinder. The lens was mounted on a lens board connected to the front standard, allowing for many of the standard movements found in view cameras. The Speed Graphic could shoot 3¼x4¼, 2¼x3¼, 4×5, and 5×7 film. Various backs could be swapped out to accept different media, including Polaroid, Fuji Quick-load, and even 120 roll film.

The Speed Graphic documented many different events for decades, making its impact on news and photojournalism difficult to measure. Some of the most iconic images from the 1930s through the 1950s were taken on Speed Graphics, including Joe Rosenthal's Pulitzer Prize–winning image of the raising of the American flag on Iwo Jima. The camera was also the camera of choice for the famous photojournalist Weegee, documenting the gritty side of New York City during that same period.

Model 95 Land Camera

In 1948 Edwin Land and Polaroid introduced the world to the first commercially available instant camera, the Model 95 Land Camera. Produced through the late 1950s, the camera went on to include two more variants, the Model 95a and Model 95b, selling over 1.5 million units.

Constructed of metal and wrapped in leather, the Model 95 was a hefty folding camera at 4.27 pounds (1.94 kilograms). The camera came with a 135mm f/11 lens equipped with a leaf shutter capable of speeds of 1/8–1/60, mounted on a handsome art deco–style metal lens board.

The Model 95 has a few interesting quirks. The most immediately apparent is that while the camera can shoot at different shutter speeds and apertures, none can be selected by the photographer. Instead, the camera has eight shooting scenarios that need to be memorized by the photographer. A dial on the lens board makes the scenario selection (1–8), viewable through a window on the lens board above the Polaroid logo. Each of the eight scenarios corresponds to a shutter speed and aperture combination, effectively making this a preprogrammed camera.

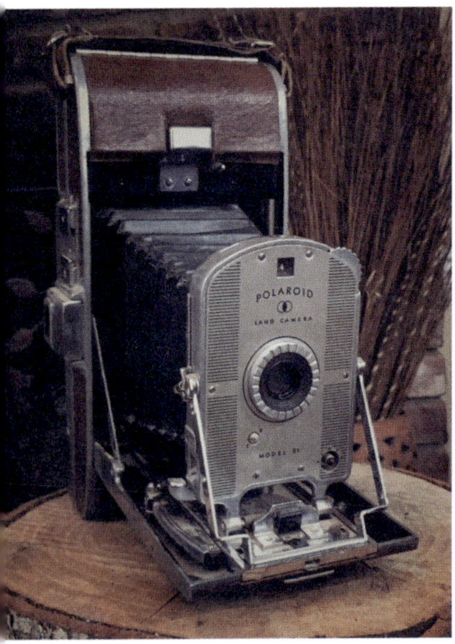

The Polaroid Land Camera can be easily identified by the lens board's handsome art deco styling and the unique single-number exposure system seen through a small viewport above the lens

Loading film requires opening three folding doors in the rear of the camera; these doors combine to create a path with a system of rollers to drag the film through, allowing it to be processed, developed, and pulled out of the camera without exposing the rest of the film to light.

The camera shot on Polaroid 40 instant roll film. The film itself consisted of both a negative and a sheet of positive paper that were eventually sandwiched together with chemicals in between by pulling a paper tab and moving the film through the complex rear of the camera. The whole process took about one minute from the start of the development to the end, leaving a 3¼×4¼-inch (8.25×10.2-centimeter) black-and-white print.

Edwin Land and Polaroid went on to simplify and speed up the process with new camera models and film, but this landmark camera introduced the world to a novel process that helped shape a new segment of photography.

Rolleiflex Automat

The Rolleiflex Automat appeared nine years after Franke & Heidecke produced the original Rolleiflex in 1928. The earlier model, a 6×6 twin-lens reflex camera, had already gone through several iterations since its debut and the brand had grown a reputation for well-constructed, well-designed cameras with extremely sharp Zeiss and Schneider lenses. The Automat represented a big step up in features, helping to set a standard on which Rollei built its next generation of TLRs.

The Rolleiflex Automat offered a wide range of improved features, though they were mostly quality-of-life upgrades: making the camera easier and much quicker to operate rather than radically redesigning an already proven formula. The Automat got its name from the upgraded mechanical wind mechanism that made many of the camera's functions

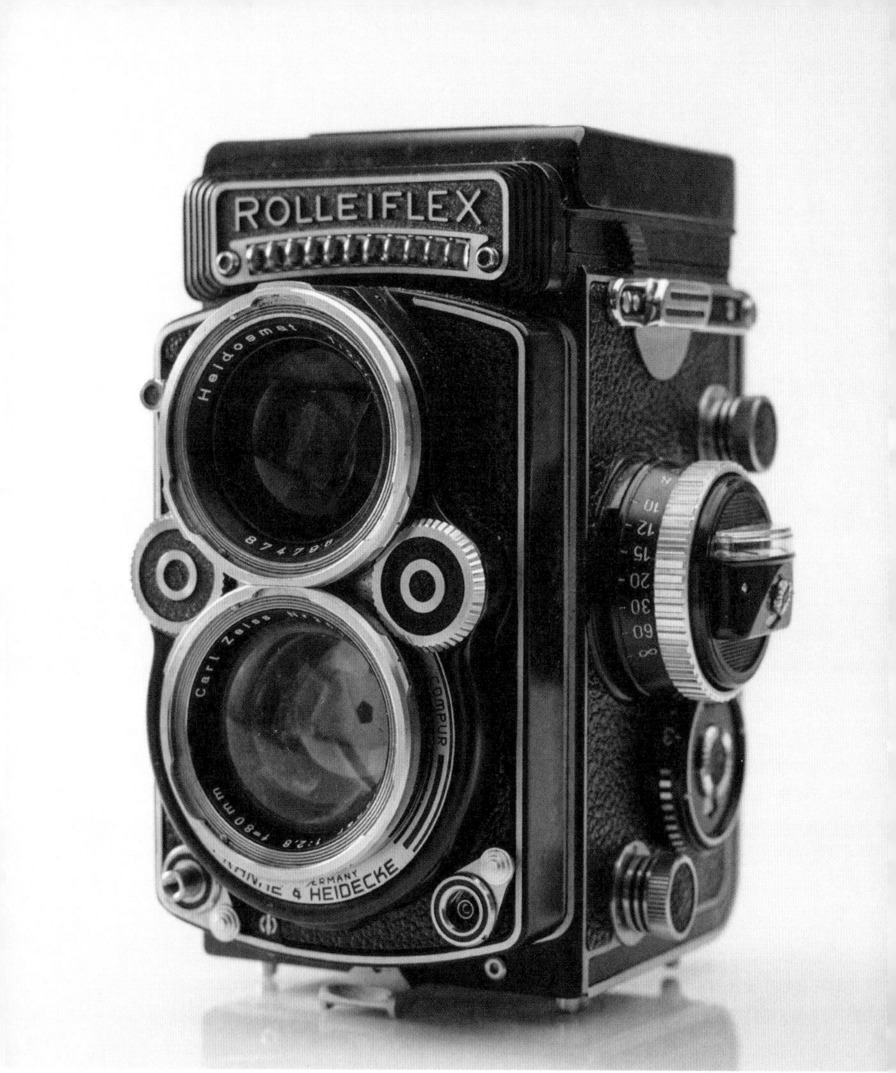

The Rolleiflex came in many different models but essentially offered two different lens speeds and focal lengths. The one pictured here is an 80mm f/2.8 lens, but the camera also came with a 75mm f/3.5 option.

more automatic. This included an easier way to load film that used a newly designed winding mechanism: A "sensor" recognized the start of the first frame based on the change in thickness as the roll transitioned from paper to film. With just a partial rotation of the winder, the film was automatically advanced to the next frame, the shutter cocked, and the frame-count window advanced. The entire process was streamlined, making the camera desirable for journalists and street photographers shooting in scenarios where speed was essential.

Being a 6×6 camera, the Automat could yield twelve shots per roll of 120 film. It came equipped with a Carl Zeiss Tessar f/3.5 taking lens with a built-in Compur-Rapid leaf shutter capable of speeds up to 1/500 and down to 1 second. The viewing lens was a Heidoscop-Anastigmat f/2.8, and the camera even came with built-in parallax correction.

Automats set a new standard for all future Rolleiflexes, and they also set the standard of excellence for what TLR cameras should be, sparking a slew of TLR cameras from other brands copying the Rolleiflex offering.

Argus C3

Some cameras are known for their smooth lines, thoughtful layouts, and beautiful industrial design. The Argus C3 is not one of those cameras. Manufactured from 1939 to 1966 in Ann Arbor, Michigan, "the Brick," as it came to be known, was a boxy, heavy, low-priced rangefinder that also happened to be one of the most successful American cameras ever produced, selling over 2.2 million units over its production run.

As its name suggests, the C3 was the third iteration in the C line of cameras from Argus, each adding new features but retaining much of their original design. Compared to the German rangefinders from Leica and Contax, the Argus is much larger, heavier, and less elegant, but by sacrificing some aesthetics for a simplified design, Argus made a camera

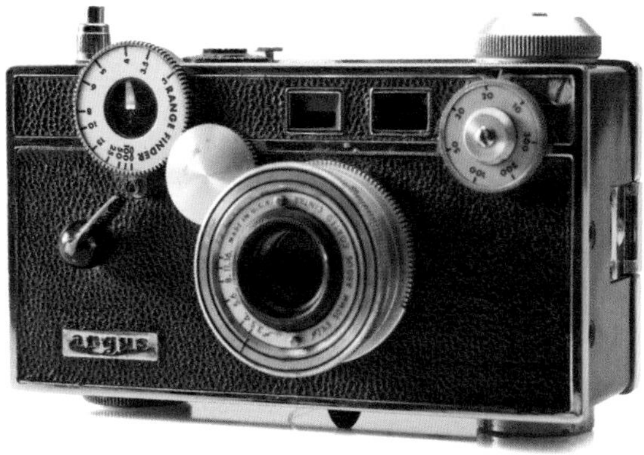

Square, boxy, utilitarian, the Argus C3 has its own charming visual characteristics, such as the way the gears that connect the focus mechanism to the rangefinder are visible on the front of the camera.

that was highly reliable at a much lower price point. The C3 originally retailed for $35, just one week's salary for the average American, but after World War II the price jumped to $78.08. Even with the price hike, though, the camera was still about a quarter of the price of a Leica III.

The C3 was constructed from a combination of Bakelite (thermoset resin) and metal covered in a leatherette exterior. The rangefinder was the coupled variety, evident on the front of the camera by the prominent gearing that connects the lens to the rangefinder knob. The camera came fitted with a 50mm f/3.5 Cintar lens that swapped out the typical four-element design for a cost-saving three-element design and was produced by a number of US optical manufacturers, including Bausch & Lomb. For the shutter, Argus opted to place a leaf shutter inside the camera body rather than inside the lens. It was capable of speeds between 1/10 and 1/300.

While the Argus C3 may not win any awards for sharpest optics or most sophisticated camera design, what the camera did offer was a well-built, reliable, and affordable camera option that made Argus the second-largest American camera manufacturer behind Kodak. They offered no-frills cameras that were built like tanks and were handed down through generations.

Hasselblad 500c

Somewhere on the barren surface of the moon sit twelve Hasselblad 500c cameras. Known for its reliability, image quality, and highly customizable modular design, the 500c rocketed into history as one of the most iconic and highly collectible film cameras available, responsible for capturing some of the most iconic images in history.

Hasselblad had been making cameras for nearly two decades before the release of the 500c in 1957. The 1000F and 1600F models that preceded it were solid performers but suffered from issues related to their focal-plane shutters. The 500c represented the first generation of a complete redesign. While the camera retained some of the core design features Hasselblads were known for—namely a 6×6 medium-format camera meant to be operated with a waist-level finder—the redesign involved replacing the focal-plane shutter with a Synchro-Compur MXV leaf shutter inside each of the lenses available for the camera. This decision made the camera mechanically simpler and more reliable, but it also came with a tradeoff: The top shutter speed was only 1/500, rather than the 1/1000 and 1/6000 found in previous models. Studio photographers weren't affected by this change, as the leaf shutter could now sync with flashes at all speeds.

A modular camera, the 500c came equipped with a newly designed 80mm f/2.8 Carl Zeiss Planar lens known for its exceptional image quality. The modular design meant that the lenses, winding crank, film backs, and viewfinder could all be swapped out for other parts and styles, allowing for a highly customizable shooting experience.

At launch, the camera with its kit lens cost $480, about $5,500 in 2025 dollars—it was not cheap. Despite its high price tag, the 500c was an incredibly popular camera for professionals, and it was soon proven in the field as a model that could hold up to the rigors of professional work while offering exceptional image quality.

The camera's popularity was taken to new heights in 1962 when astronaut Walter Schirra convinced NASA to use the 500c on the Mercury space mission. NASA obliged, and several highly modified cameras (stripped of their leather covering, auxiliary shutter, reflex mirror, and

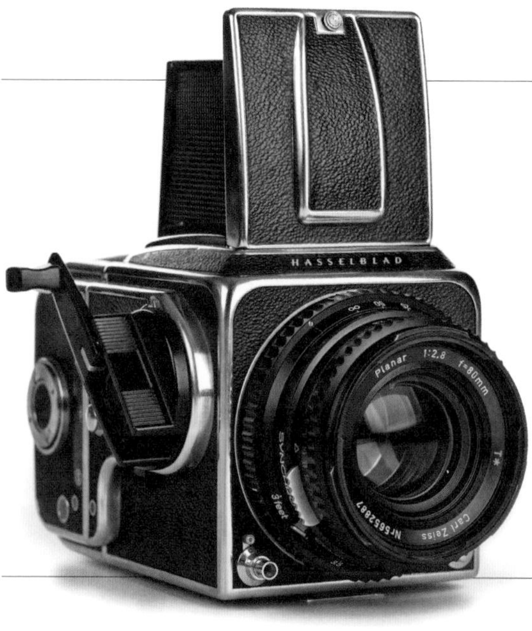

The 500c saw several variants over the years, each adding features. The 500c/m (seen here), launched in 1970, added interchangeable focus screens.

viewfinders) were taken into space aboard Mercury 8 in October 1962. This marked the first of several NASA missions with Hasselblads, including the Apollo moon landing seven years later.

The camera became the preferred medium-format camera of many famous photographers, such as Sid Avery and Ansel Adams. It also served as a blueprint for other manufacturers (among them Rollei, Mamiya, and Bronica) to replicate, but none ever achieved the same quality as the original.

Polaroid SX-70

In 1972, while standing onstage at the annual company meeting, Polaroid founder Edwin Land reached into his suit jacket pocket and pulled out a flat, thin, metallic object. Within seconds the object popped up into a fully functioning camera and produced five physical photographs, a feat that no Polaroid Land cameras had ever performed before.

This was the debut of the Polaroid SX-70, a collapsible, folding SLR featuring a brand-new film technology from Polaroid: Zero-Time film, an instant film capable of vibrant colors and quick developing, containing all the necessary chemicals within a sealed package that no longer required peeling apart or specific timing by the photographer. It was a new era for Polaroid.

The SX-70 was handsome, some might even say cute, when the camera was unfolded. Construction was plastic plated in a metallic mixture of copper-nickel-chromium alloy with leatherette accents. As a foldable SLR, the camera was designed with a relatively complex set of three aspheric mirrors positioned at odd angles to direct light through the camera and up into the viewfinder while also flipping the image right-side up for the photographer. A 116mm f/8 lens was coupled with an electronic shutter, both of which were connected to a cadmium

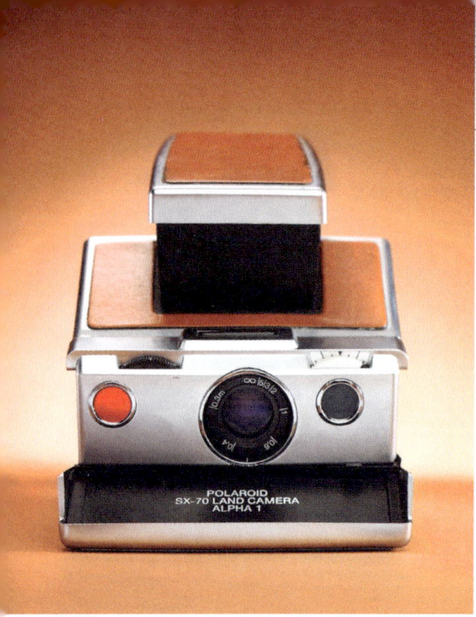

Chrome accents, leather trim—the SX-70 is clearly a chic camera, especially when folded up. Later models featured a large circular disk on the face of the camera for its sonar autofocus.

sulfide (CdS) exposure meter capable of full programmed automatic exposure.

The SX-70 was the first camera to introduce Polaroid pack film. Unlike earlier Land cameras, whose rolls of film had to be manually pulled through the camera with paper tabs, the SX-70's film came in plastic cartridges with ten exposures. The cartridge was loaded into the base of the camera, which also housed rollers to squeeze and spread chemicals across the film and motors to drive those rollers, ejecting the photo out the front of the camera. One of the more interesting design decisions was to place the battery not in the camera itself but in the film packs: Every time you loaded a new film cartridge, a fresh battery was loaded as well.

The camera went through many upgrades through the years, including a big one in 1978 with the introduction of sonar autofocus, the very first SLR camera to come with an active autofocus system.

The design and film technology eventually gave way to what most people think of when they think of Polaroids: Polaroid 600 film and cameras, the company's most commercially successful and accessible format. But even with the evolving film format and more advanced cameras to come, the SX-70 remains one of most visually iconic cameras ever produced by the company.

Nikon F

When Nikon debuted the Nikon F in 1959, it was impossible to predict how profound an impact it would have on the company and, to a larger extent, the camera world. Nikon had been producing cameras for eleven years by this point, but everything in their catalog had been rangefinders. Manufacturers in the mid-1950s believed that SLRs provided more advantages over rangefinders than disadvantages, so in 1957 Nikon began development of the Nikon F, their first SLR.

Nikon had a number of goals in mind: They wanted the camera to be as easy and quick to use as their rangefinders; they wanted to design a "system camera" that was modular enough to accommodate a large system of accessories; the lens mount needed to handle a variety of lenses that could work with an automatic mirror mechanism and not produce heavy vignetting; and the whole system needed to be extremely robust and reliable. By all accounts, they nailed it.

Hasselblads weren't the only cameras to be used by NASA: Modified Nikon F series were also found on Apollo missions over the years.

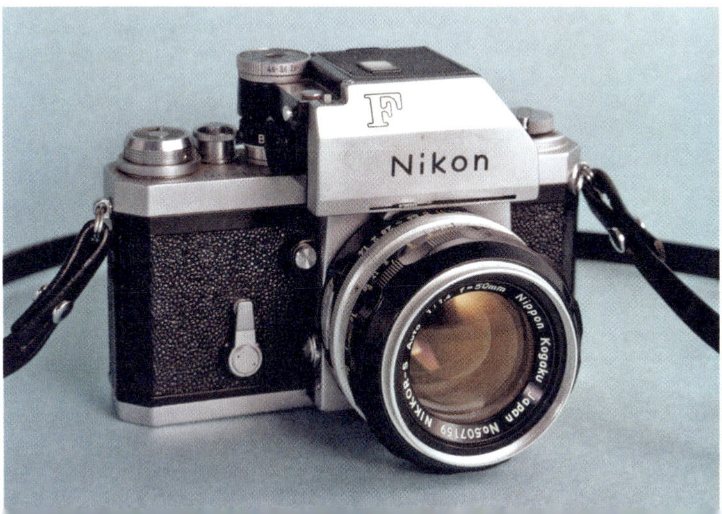

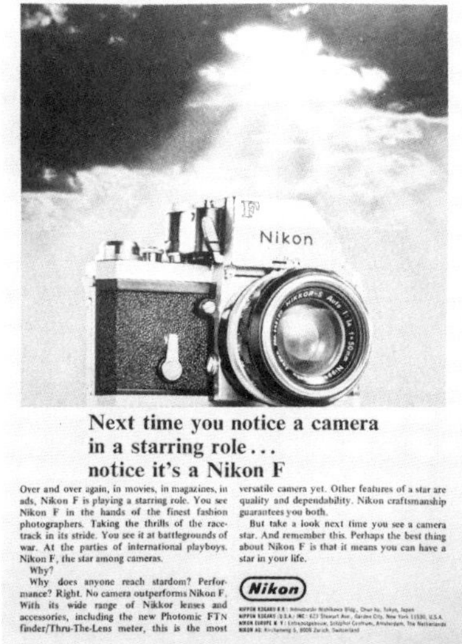

The Nikon F gained a reputation for being so robust and reliable that it was lovingly referred to as a "hockey puck," something that can take a beating and keep on working.

When the Nikon F came to market in 1959, it had an eye-level viewfinder with a split focusing screen, both of which could be swapped for other options like a waist-level finder or one of many different focusing screens. The mirror had an automatic return, removing the need to cock the shutter to drop the mirror down as on other SLRs. The shutter was made from titanium foil. The camera was tested in extreme temperatures and conditions to ensure the mirror and shutter mechanisms could perform at least 100,000 cycles while also withstanding harsh environments. And the camera came with a new F lens mount that would be used by Nikon for over forty years, even after they made the transition to digital.

The Nikon F was by all metrics a massive success. Due to its reliability, advanced features, and ever-growing range of lenses and accessories, it was the first 35mm SLR that was widely adopted by professionals. Used everywhere from the Olympics to the battlefields of the Vietnam War—and even on the International Space Station—the Nikon F set the standard for what a professional 35mm SLR should be. It also marked the start of the F-series line, which remained in production through 2020 with the F6, the last production model.

Canon AE-1

The release of the AE-1 in 1976 marked a new era for Canon, showing the company and the photo world what was possible when you integrate microprocessors deep in the heart of a camera. While the AE-1 was not the first SLR to incorporate a microprocessor (that award goes to the Nikon Nikkormat EL four years earlier), it *was* built from the ground up around the microprocessor unit. Handing over more functions to the processor enabled Canon to reduce the number of internal parts needed by three hundred, thereby allowing more production automation and leading to a more advanced and less expensive camera for the masses. And with 5.7 million units sold, the masses certainly took to it.

While the AE-1 was in development, Canon found itself losing a lot of ground in the professional camera space to Nikon, whose F line was dominating the SLR market among working professionals. Instead of directly competing with that line, Canon targeted amateurs. The microprocessor unit enabled the camera to electronically control metering and exposure calculations, set the aperture value, and even control shutter operations via an electromagnetic shutter.

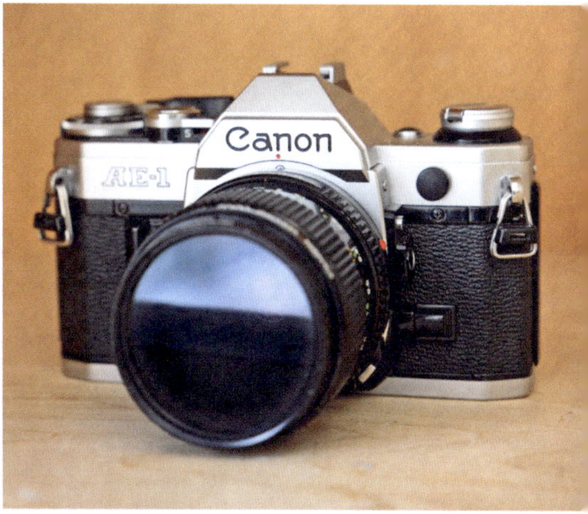

While a deep integration of electronics brought many advantages to cameras like the AE-1, the camera marked a new age with an electronic shutter that wouldn't fire without batteries.

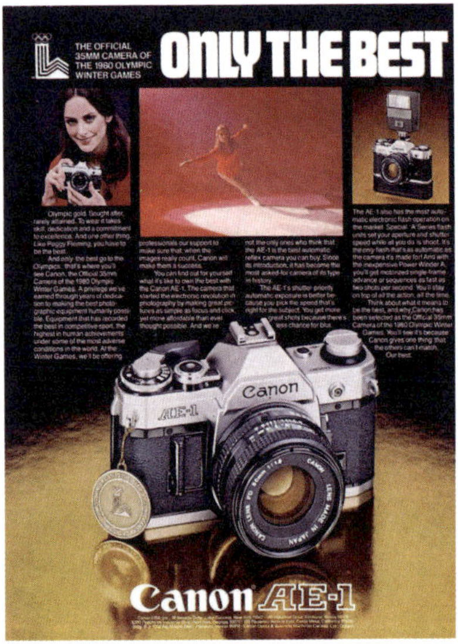

As the official camera of the 1980 Winter Olympics, the Canon AE-1 allowed Canon to make inroads with professional sports photographers.

The initial feeling among many professionals was disdain, claiming the camera took away the magic of decision-making that photography was all about. As it turns out, though, these were exactly the features that spoke to amateur photographers looking for a low barrier to entry into photography.

The AE-1 was built as a shutter-priority camera with a full manual mode. This meant that, with the lens aperture set to auto, the photographer could set the shutter speed and the camera would use the light meter readings to automatically set the aperture. In manual mode, the photographer would set the desired shutter speed and the light meter inside the camera would then tell them what aperture to select.

The AE-1 holds the record for the best-selling film SLR ever produced, but its real impact can be felt in its influence on Canon's later camera-making decisions. Not only did the camera spur a whole line of microprocessor-driven A-series cameras, it showed the company that computer-driven cameras were the future, eventually leading Canon to develop the EOS system that made it the largest camera company in the world.

Leica M3

Considered by many to be the perfect rangefinder, Leica's M3 embodies much of what makes mechanical film cameras so enticing: exceptional build quality, precision engineering, sharp optics, and timeless design. Leica is considered the company that set the standard for rangefinder cameras, and that is in no small part thanks to the M3.

Leica had been making cameras for three decades when it released the M3 in 1954. The company had produced several submodels since the Leica I in 1925, but those cameras all represented incremental updates. The M3, however, marked the beginning of a completely new generation of cameras for Leica. Everything about the M3 made it quicker and easier to use than its predecessors while still retaining its fully mechanical design and high build quality.

The M3 introduced a new bayonet-style lens mount that replaced the screw mounts found on previous Leica cameras. This new "M-mount" allowed much faster lens changing. It also allowed the frame lines within the viewfinder to change dynamically based on the focal length of Leica's 50mm, 90mm, and 135mm lenses. Speaking of

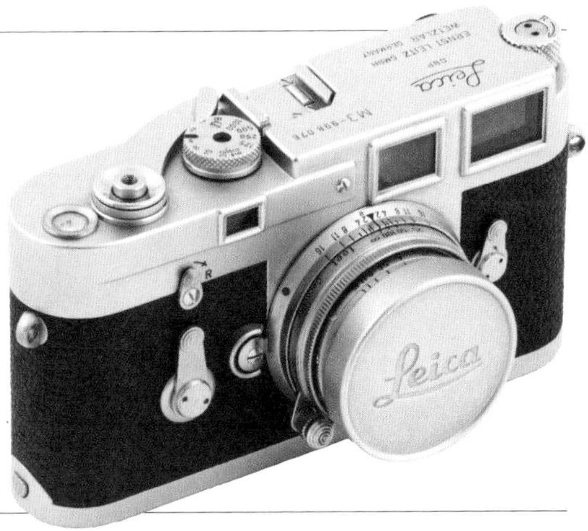

The first editions of the M3 contained a film-advance lever that had to be pulled twice because Leica was nervous about tearing the film. It was later replaced with a single-stroke lever.

the viewfinder, unlike the previous models that had separate view- and rangefinder windows, the M3 combined the two into one large, bright window that included parallax correction.

Loading and changing film also got easier thanks to a hinged camera back. Though film replacement still required the bottom plate to be removed, the center of the back of the camera could now swing up. A double-stroke film-advance lever replaced the advance knobs of the Leica III—later replaced by a single-stroke lever—and the frame counter automatically reset when a new roll of film was loaded. All the necessary adjustments could now be seen from the top of the camera, including a newly designed shutter-speed dial that combined fast and slow shutter-speed selection in a single dial, a welcome upgrade from the two dials found on previous models.

During its production run, the M3 sold about a quarter of a million units, making it one of the most successful models in the Leica M series and the preferred camera of a number of famous photographers, including Nick Ut, Elliot Erwitt, Henry Cartier-Bresson, and even the Queen of England.

Rectaflex

Rectaflex was established in Rome in 1948 by lawyer and industrialist Telemaco Corsi. The company was only in existence for a short time, going out of business in 1955 (only to be resurrected by the Prince of Lichtenstein for a few years and then permanently closed in 1958). Although short-lived, the company cemented its place in camera history with the release of the Rectaflex camera in 1948.

The Rectaflex was the first (and only) Italian-made SLR and, more significantly, the first production SLR to contain a pentaprism, though Contax began researching the feature before World War II (more on that in the next section).

The Rectaflex was a 35mm SLR with a focal-plane shutter capable of speeds from 1 second to 1/1000. It had a wide-diameter bayonet-style lens mount and a mirror design that operated somewhere between the instant-return mirrors of the 1950s and the manual-return mirrors of earlier cameras. Rectaflex's "rapid-return mirror" worked by coupling the pressure placed on the shutter button with the mirror mechanism. As you pressed down on the shutter, the mirror flipped up (press harder or faster and the mirror moved faster); as pressure was reduced, the mirror was lowered. This interesting design feature was more convenient than having the mirror tied to cocking the shutter, but not as convenient as the instant-return mirrors that would come later.

The real star feature of the camera was the eye-level finder with a pentaprism. While there certainly were concepts and examples of cameras with prisms in them before this, none were put into commercial production before the Rectaflex. The prism allowed the image to be flipped horizontally and vertically, allowing the camera operator to see the image right-side up through the viewfinder. The camera came with a split focusing screen and, according to one magazine advertisement, was "the only 35mm camera featuring the unique duofocus system."

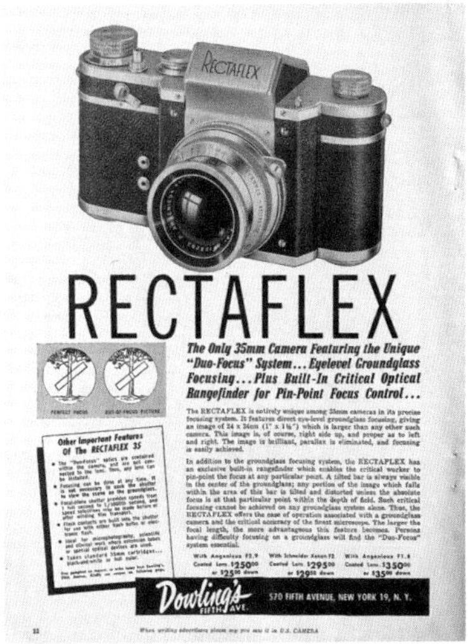

As seen in the ad, the Rectaflex came equipped with three different lens models, each set at a different price. The cheapest started at $250.

The camera certainly is handsome, designed with a

nice mixture of hard geometric lines and rounded features. The top of the camera has a uniquely shaped "mound" housing the pentaprism and is topped with an attractive art deco "Rectaflex" wordmark. While Rectaflex certainly wasn't a brand that persisted, the camera introduced a few interesting design features that wound up being adopted by other companies. The cameras themselves remain quite collectible today.

Contax S

In the late 1930s, Zeiss Ikon started developing plans for an SLR with an eye-level viewfinder and a pentaprism. World War II put a hold on production, which began again in the fall of 1949 when VEB Zeiss Ikon started manufacturing the Contax S.

The Contax S was a 35mm SLR that *just* missed out on the title of "world's first SLR with a pentaprism" to the Rectaflex, which went into production the year before. There is little doubt, however, that Contax had been working on the pentaprism design long before other companies, and might have produced its version earlier had the war not intervened.

The camera included a 42mm screw-on lens mount introduced by Praktiflex a few years earlier, a design that later became the standard mount type for many camera manufacturers. An excellent 58mm f/2 Carl Zeiss Jena Biotar lens was included with the camera, mounted in front of a cloth focal-plane shutter capable of speeds between 1 second and 1/1000. Advertisements at the time called it the "First eye-level automatic reflex camera referring to the "automatic" nature of the film knob's ability to advance the film, cock the shutter, move the exposure counter, and prevent double exposure.

After World War II, Germany was divided. East Germany rebuilt many of the camera-manufacturing companies in Dresden, a city that

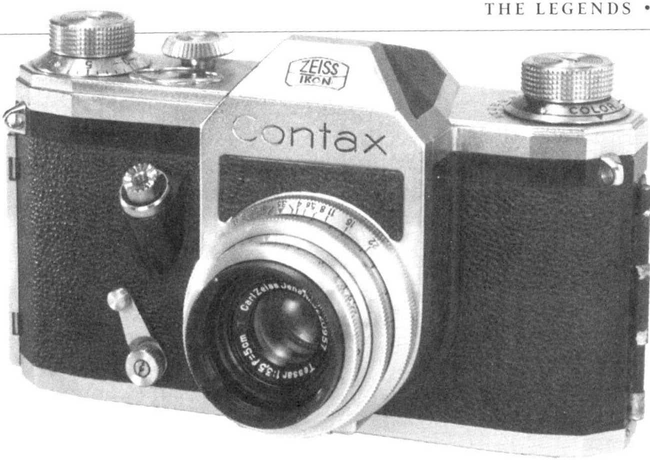

While not a rangefinder, when comparing the Contax S with the Leica 1, it's hard not see the design inspirations taken from Leica—though the Contax featured a unique shutter button placement.

had sustained catastrophic damage during the war. It was here that VEB Zeiss Ikon released the Contax S. Unfortunately, the Contax S was plagued by supply shortages, manufacturing problems, and poor materials. The first run of cameras lasted only a couple of months before the factory shut down to address manufacturing issues.

Under pressure from the West German Zeiss Ikon AG to stop using their name, VEB Zeiss Ikon eventually abandoned the use of Zeiss Ikon and Contax brand names, and the Contax S transformed slowly into the Pentacon, a combination of pentaprism and Contax. The camera never quite caught on after the name change, and production finally ceased. Interestingly, the company also held the rights to the name Pentax (another combination of the words "pentaprism" and Contax); they sold the rights to the Japanese camera manufacturer Asahi in 1957, a company that eventually rebranded itself as Pentax.

Olympus OM-1

When Olympus introduced the OM-1 in 1972, the market was full of large, heavy, chunky SLRs that were so big and rugged they were even able to stop bullets. Just ask photojournalist Don McCullin, whose Nikon F1 stopped an AK-47 bullet and saved his life. While that may be great for professionals, Olympus recognized that there was a market for a smaller, lighter, but still robust camera. And it just so happened that designing smaller, lighter, and robust equipment was a specialty of Olympus.

Enter the OM-1. It was designed by a team led by legendary designer Maitani Yoshihisa, the designer of the Olympus Pen cameras. When released in 1972, the OM-1 held the record for the smallest and lightest SLR in the world.

To accomplish their design goals, Olympus had to go through the camera and redesign many of its standard internal components. The shutter, typically made of thin metal blades or rubberized cloth, was

The OM-1 came in many versions over the years, such as the OM-1MD, released in 1970, which allowed for the attachment of a motor drive but kept its same base design.

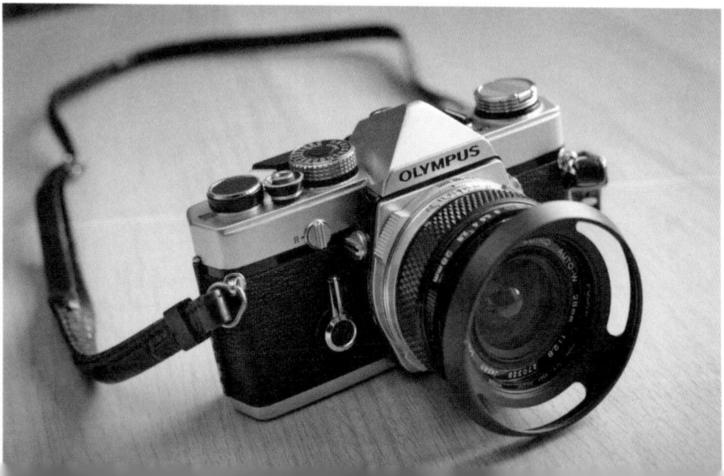

replaced by strings. A new air-dampening system reduced the vibration of the mirror stop, and the whole system was stress-tested to ensure the shutter and mirror mechanisms were capable of at least 100,000 actuations. The focusing screens were interchangeable and the camera was capable of TTL metering.

At launch the camera was named the M-1, but it was so successful that it drew the ire of Leica, whose flagship line of rangefinders already held the M moniker. After an initial production run of about 52,000 cameras, Olympus agreed to rename the camera line, changing it to the OM-1, an abbreviation of Olympus Maitani, named for the chief designer at Olympus. The OM line was a smash hit, offering a range from professional cameras to amateur cameras for every kind of consumer. Olympus produced nineteen models between 1972 and 2002, and the OM designation remains on their digital cameras today.

Pentax Spotmatic

When Asahi Pentax debuted the Spotmatic at Photokina in 1960, it was the world's first SLR with TTL metering. The system was designed with a small arm containing a light-measuring system on the end that sat in the middle of the viewing area. The system effectively acted like a spot meter, which is how the camera got its name. That system, however, was scrapped after testing determined that an average metering system (which metered the entire scene) was more useful to the everyday consumer. The Spotmatic kept its name by the time the camera reached public hands in 1964, but it had swapped the metering design for an average metering system driven by two CdS cells, one on each side of the viewfinder.

The Spotmatic was a completely new camera design for Asahi Pentax, from the inside out. The shutter had been overhauled, now capable of at least 100,000 actuations, which was a massive improvement from

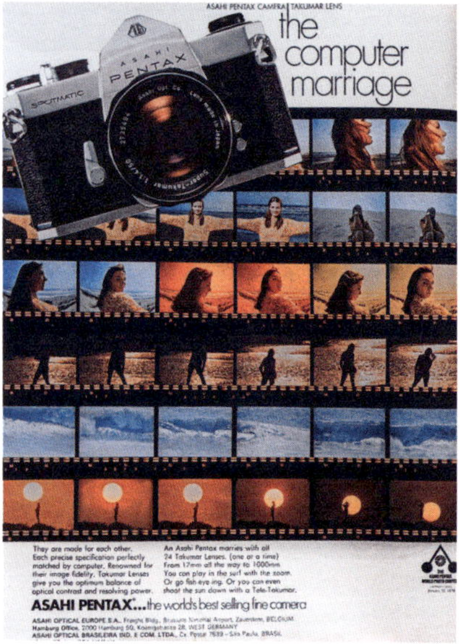

The Spotmatic's advertising often featured its advanced technology, as in this ad celebrating the computer technology used to make its precision lens.

the 20,000 rating of their previous cameras. The mirror system was redesigned to be more robust, the camera body and mirror box were machined out of aluminum, and the film transport system (rails and pressure plate) was completely redesigned to keep the film flatter while shooting. And the camera came fitted with an M42 screw-on lens mount, opening the camera to a large selection of high-quality glass not only from Pentax (Takamar) but also companies like Carl Zeiss.

Taking a meter reading required the photographer to stop down the lens to the desired aperture and slide a button up on the side of the camera. This set of actions, while revolutionary at the time, proved a little cumbersome and eventually was replaced with a TTL design that did not require stop-down metering. The first camera to offer this new design from Pentax was the Electro-Spotmatic/ES, released in 1971.

The Spotmatic would launch an entire line of cameras with that name from 1964 until production ended in 1977, selling an impressive 4 million units.

Minolta Maxxum 7000

The Minolta Maxxum 7000 secures its place in the history books as the first camera to incorporate autofocus *and* a film-advance motor drive. While other cameras had one or the other of those features, none had incorporated both. In time this design became standard in electronic SLRs.

With its boxy, squared-off angles, button-driven settings, and liberal use of black plastic, the Maxxum 7000's design is emblematic of the 1980s in which it was released. Outside of the breakthrough design that located autofocus and electronic film advance inside the camera, the camera itself offered features that were pretty much standard at the time. Gone are any manual, mechanically driven controls, replaced now by electronically selectable controls.

The Maxxum 7000's styling is emblematic of the period in which it was built, featuring hard, boxy lines, copious amounts of plastic, and LCD screens.

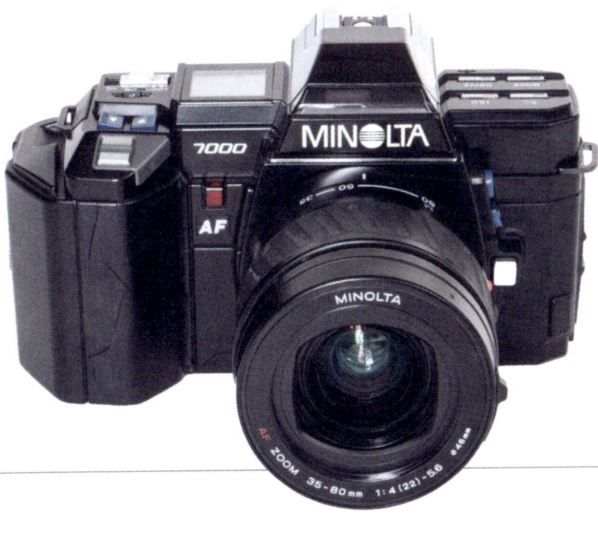

Like Canon with its introduction of the EF mount, Minolta abandoned previous mount designs with the Maxxum 7000 and introduced the new A-mount system, breaking compatibility with old lenses. The new camera mount and design allowed Minolta to move the autofocus sensors and focusing drive inside the camera, however, in turn allowing for lenses to become smaller and cheaper to produce. Interestingly, the A-mount system lives on today through Sony, which acquired Minolta's assets in 2006 and continued the use of the mount in its A-series cameras.

Pentax K1000

The inclusion of the Pentax K1000 is more of an ode to a beloved classic camera than it is calling out a landmark camera that stands out in terms of breakthrough features or technology. Produced by Asahi Pentax from 1976 to 1997, the Pentax K1000 is known as a reliable, no-frills, inexpensive camera that found itself in many classrooms and in the hands of millions of photographers around the world, largely due to its blend of affordability and reliability. Selling more than 3.5 million units over its run, the K1000 became the single best-selling camera in Pentax's lineup and one of the most successful models in history.

The K1000 is nearly all metal, with fully mechanical internals and fully manual lenses. The camera had no electronic parts that could go wrong, making the camera fairly easy to maintain. The one part that did require some power was the light meter, which operated off a small silver oxide battery (still readily available today); however, the camera doesn't rely on a functioning light meter to work.

The camera came with far fewer features than nearly all other SLRs at the time, but what it lacked in features it made up for in affordability. The K1000 was the simplest camera in Pentax's K-series cameras, stripped of any extras like a self-timer or depth-of-field preview button.

The viewfinder lacked any information about current camera settings (aperture and shutter-speed indicators were absent) and instead contained just a simple light meter. This made the camera less desirable for those who wanted assistance with some of the operations, but it was appealing as a student camera for instructors teaching photography, since it forced the operator to understand how to manually control all aspects of camera operation. The price didn't hurt either: The camera launched at a suggested retail price of $299.50 with a SMC Pentax 55mm f/2 lens (and it often sold far below that price), making it quite an affordable package.

K1000 naming conventions followed a similar scheme to those used in other Pentax models. As witnessed in some of the Spotmatic models (SP500 and SP1000), Pentax incorporated the maximum shutter speed into the name of the camera. Thus, the K1000 had a maximum shutter speed of, you guessed it, 1/1000.

While the K1000 may not have revolutionized the technology within a camera, it did offer an affordable, reliable workhorse that found its way into the hands of many and is still beloved today.

The K1000 was produced in China and Japan. The Japanese models can be identified by a nameplate that includes "Asahi," while the Chinese versions only say "Pentax."

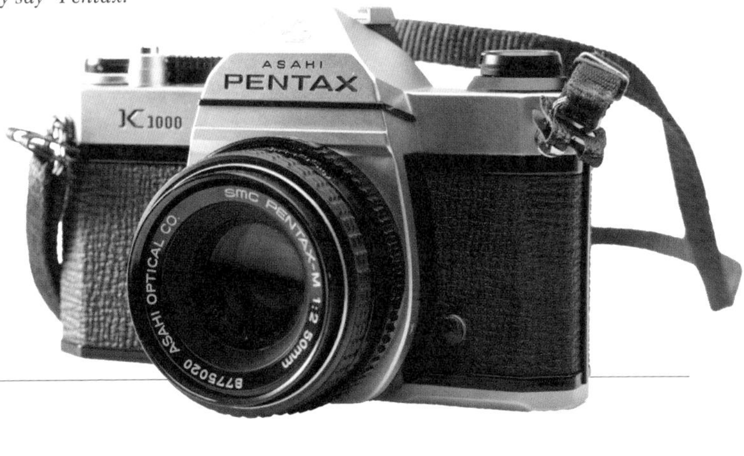

About the Author

Aaron Lurth an educator, photographer, filmmaker, and designer whose photographic work has been used by *National Geographic*, ABC, NBC, CBS, MTV.com, *Time*, *The Des Moines Register*, CNET.com, Red Cross, and *The Iowa Review*. His film projects have screened at over 100 festivals worldwide and have won a number of awards, including the Iowa Motion Picture Association's Best Documentary film. He currently serves as assistant professor of art and visual communication at Luther College in Decorah, Iowa.

Photo Credits

Alamy Stock Photos: 4 (Vaidas Bucys), 7 (Oxford Science Archive/Heritage Images), 10 (The History Collection), 12 (GL Archive), 13 (INTERFOTO), 15 (Oxford Science Archive/Heritage Images), 16 (Patti McConville), 17 (Historic Collection), 18 (Paul Gorvett), 19 (Picture Partners), 21 (D. CHIARELLI), 23 (Tetiana Lazunova), 24 (James Jackson), 25 top (Andreas Gerhardinger), 25 middle (Ian Shipley TYS), 27 (gameover), 28 (f28 Media), 29 (theLIME), 31 (Alexander Staridas), 33 (INTERFOTO), 35 (Michelle Bridges), 37 (Ken Hawkins), 38 (Rene Pirolt), 45 (Jonathan Eastland), 46 (BrinkBox), 49 (Felix Choo), 51 (PBWPIX), 52 (Patti McConville), 54 (Shawshots), 55 (UrbanImages), 56 (Photology1971), 57 (William Scott), 58 (Retro AdArchives), 60–61 bottom (History and Art Collection), 61 top (John Gilbey), 63 (Shawshots), 66 (Daniel Vrabec), 69 left (Shawshots), 69 right (Peregrine), 71 (adsR), 72 (Chris Hellier), 75 (delfimsaneiva), 76 (Patti McConville), 78 (Stuart Forster), 80 (Alberto Maisto), 81 (Anil Dave), 85 (Patti McConville), 86 (flab), 89 (Shawshots), 91 (Circa Images), 95 (Uwe Deffner), 96 (Shawshots), 97 (Anil Dave), 99 (Kazimierz Jurewicz), 100 (Kelly Tippett/Stockimo), 106 (Jack Grieve), 108 (Oliver Leedham), 109 (H. Armstrong Roberts), 110 (adsR), 112 (adsR), 113 (Jeff Morgan 06), 118 (Thousand Word Media), 120 (benarchival), 121 (Stoxo), 125 (H. Armstrong Roberts)

Author Photos: 32, 40, 42, 47, 48, 65, 73, 83, 102, 104, 111, 115, 123

Getty Images: 92 (Science & Society Picture Library), 94 (Science & Society Picture Library), 117 (Science & Society Picture Library),

Library of Congress: 8 (Carol M. Highsmith, Ref. No. LC-DIG-highsm-20822)

Index

Quarto.com

First Published in 2026 by Motorbooks, an imprint of The Quarto Group,
100 Cummings Center, Suite 265-D, Beverly, MA 01915, USA.
T (978) 282-9590 F (978) 283-2742

EEA Representation, WTS Tax d.o.o.,
Žanova ulica 3, 4000 Kranj, Slovenia.
www.wts-tax.si

Motorbooks titles are also available at discount for retail, wholesale, promotional, and bulk purchase. For details, contact the Special Sales Manager by email at specialsales@quarto.com or by mail at The Quarto Group, Attn: Special Sales Manager, 100 Cummings Center, Suite 265-D, Beverly, MA 01915, USA.

30 29 28 27 26 1 2 3 4 5

ISBN: 978-1-57715-676-5

Digital edition published in 2026
eISBN: 978-1-5771-5677-2

Library of Congress Cataloging-in-Publication Data available

Cover Illustration: Sue Boylan

Printed in Guangdong, China TT032026